URBAN CHANNEL DESIGN AND FLOW ANALYSIS

URBAN CHANNEL DESIGN AND FLOW ANALYSIS

By

James C.Y. Guo

and

Yuan Guo

Water Resources Publications, LLC Since 1971 WRP Colorado

For Information and Correspondence:
Water Resources Publications, LLC
P. O. Box 630026, Highlands Ranch, Colorado 80163-0026, USA

URBAN CHANNEL DESIGN AND FLOW ANALYSIS

By

James C.Y. Guo, PhD

And

Yuan Guo

ISBN-13: 978-1-887201-92-6

Library of Congress Control Number: 2017964637

This publication is printed and bound in the United States of America.

FOREWORD

Urban channel *is* one of the major components in the urban infrastructure systems. Channels are built to alleviate flood damage and to remove surface runoff away from the urban area. Today, urban channels shall be designed with composite cross sections to cope with multiple storm events, and landscaped with recreational features as a natural resource of open space in an urban area. In this book, the subject of urban channel design is presented to include not only the protocols for hydraulic procedure, but also concerns of public safety, esthetics as greenbelt, and economic consideration between costs and benefits.

Water flow in a channel is driven by the gravity. A water flow presents its duality between the flow depth and kinetic energy. On a steep slope, a hydraulic jump may be triggered when the downstream tailwater is sufficient while a subcritical flow results in a hydraulic drop into a pool. In this book, both hydraulic jump and drop are reviewed in theory and then discussed with numerous examples on the designs of plunging pools for energy dissipation. Of course, the theoretical approaches are further compared and examined with empirical formulas for engineering practices.

As always, there are tradeoffs in the objectives when designing a channel. To increase the channel hydraulic capacity, a steep gradient is preferred. On the other hand, the long-term maintenance due to streambed erosion and deposition demands a mild slope for channel flows. In this book, the method of drop structure is presented as a solution to reduce the channel's slope. For comparison, the riprap protection is also presented in case that a higher hydraulic capacity is preferred.

This book applies the green concept to urban channel designs. All channels shall be designed with considerations of wetlands, greenbelts, water-front parks, and recreational open space with bike paths and sport fields. All hydraulic structures across the floodplain need to achieve the hydraulic goals, and also to accommodate recreational purposes such as rafting, kayaking, and wild life habitats. As always, urban channel shall be designed to be the assets to better the urban life, and as the major infrastructure to improves the quality of water environment.

James C.Y. Guo, PhD, P.E. Professor/Director
University of Colorado Denver, Denver, Colorado

TABLE OF CONTENTS

Foreword v

List of Figures x

List of Tables xii

CHAPTER 1 - OPEN-CHANNEL HYDRAULICS 1

1.1 CLASSIFICATION OF CHANNELS 1
1.2. CLASSIFICATION OF CHANNEL FLOWS 2
1.3. SLOPES IN CHANNEL FLOW 3
1.4 CROSS SECTIONAL ELEMENTS 5
1.5. EMPIRICAL FORMULA 7
1.6. ROUGHNESS COEFFICENT 9
1.7. NORMAL FLOW 10
1.8. CLOSING 14
1.9. HOMEWORK 15

CHAPTER 2 - HYDRAULIC DROP AND JUMP 17

2.1. SPECIFIC ENERGY 18
2.1.1. Critical Flow on Specific Energy Curve 21
2.1.2. Special Case for Specific Energy 21
2.2. SPECIFIC FORCE 23
2.2.1. Critical Flow on Specific Force Curve 28
2.2.2. Special Case for Specific Force Curve 28
2.3. CLOSING 29
2.4. HOMEWORK 30

CHAPTER 3 - ENERGY DISSIPATION BASIN 31

3.1. WEIR HYDRAULICS 31
3.1.1. Rectangular Weir 31
3.1.2. Triangular Weir 32
3.1.3. Trapezoidal Weir 33
3.2. STILLING POOL 34
3.2.1. Design Flow 35
3.2.2. Operational Flow 37
3.3. PLUNGING POOL 39
3.3.1. Low-Flow Weir on Top of Drop Structure 41
3.4. HOMEWORK 42

CHAPTER 4 - DROP STRUCTURE FOR GRADE CONTROL 45
4.1. CONCEPT OF PERMISSIBLE VELOCITY 45
4.2. CONCEPT OF PERMISSIBLE FROUDE NUMBER 48
4.3. DESIGN CONSIDERATION 49
4.4. PLUNGING POOL 52
4.5. WATER SURFACE PROFILE 56
4.7. HOMEWORK 59
CHAPTER 5 - CONCRETE CHANNEL 61
5.1. DESIGN CONCERNS IN CONCRETE CHANNEL 61
5.2. HEIGHT OF FREEBOARD 65
5.3. SUPERELEVATION 66
5.4. EFFICIENT CHANNEL SECTIONS 68
5.4.1. Efficient Trapezoidal Channel without Freeboard 71
5.4.2. Efficient Rectangular Channel with Freeboard 72
5.4.3. Efficient Rectangular Channel with No Freeboard 73
5.5. CLOSING 76
5.6. HOMEWORK 77
CHAPTER 6 - GRASS CHANNEL 79
6.1. GRASS LININGS 80
6.2. MANNING'S ROUGHNESS FOR GRASS LININGS 82
6.3. GRASS CHANNEL DESIGN 83
6.4. CLOSING 86
6.5. HOMEWORK 89
CHAPTER 7 - RIPRAP CHANNEL 91
7.1. RIPRAP LINING 91
7.2. DESIGN CONSIDERATIONS 92
7.3. STREAM POWER-BASED METHOD 92
7.4. SHEAR STRESS-BASED METHOD 96
7.4.1. Roughness Coefficient 96
7.4.2. Flow Shear Stress 97
7.4.3. Permissible Shear Stress 101
7.5. CLOSING 103
7.6. HOMEWORK 104
CHAPTER 8 - FLOOD CHANNEL DESIGN 105
8.1. LOW FLOW CHANNEL 107
8.2. TRICKLE CHANNEL 108
8.3. WETLAND CHANNEL 109
8.4. DESIGN CRITERIA FOR COMPOSITE CHANNEL 110

8.5. CONVEYANCE CAPACITY IN COMPOSITE CHANNEL ... 111
8.5.1. Flow in Main Channel ... 111
8.5.2. Flow in Overbank Areas ... 112
8.6. CLOSING ... 117
8.7. HOMEWORK ... 118
CHAPTER 9 - HIGH GRADIENT CHANNEL ... 121
9.1. STABILITY OF SUPERCRITICAL FLOW ... 122
9.2. ROLL WAVES ... 126
9.3. CLOSING ... 129
9.4. HOMEWORK ... 130
CHAPTER 10 - CULVERT HYDRAULICS ... 131
10.1. CULVERT DESIGN CONSIDERATIONS ... 132
10.2. CULVERT SIZING ... 134
10.3. CULVERT HYDRAULICS ... 137
10.3.1. Culvert Hydraulics under Inlet Control ... 138
10.3.2. Culvert Hydraulics under Outlet Control ... 139
10.3.3. Determination of Culvert Capacity ... 141
10.4. CLOSING ... 142
10.5. HOMEWORK ... 143
REFERENCES ... 145

List of Figures

Figure 1.1. Natural and Man-made Channels. 1
Figure 1.2. Classifications of Channel Flows. 3
Figure 1.3. EGL and HGL in Channel Flow 4
Figure 1.4. Channel Alignment Between Points. 5
Figure 1.5. Cross Sectional Elements in Trapezoidal Channel 6
Figure 1.6. Water Flow in Concrete Box. 8
Figure 1.7. Uniform and Non-uniform Flows. 10
Figure 1.8. Flow Parameters for Example. 11
Figure 1.9. Example Rating Curve. 12
Figure 1.10. Rating Curve at Stream Gage. 14
Figure 1.11. Stage-Flow Measurement. 15
Figure HW1.1. Slopes on HGL and EGL. 15
Figure 2.1. Hydraulic Jump and Drop. 17
Figure 2.2. Specific Curve. 19
Figure 2.3. Application of Specific Energy Curve. 20
Figure 2.4. Calculating Specific Force. 24
Figure 2.5. Specific Force Curve. 25
Figure 2.6. Application of Specific Force Curve to Jump. 26
Figure 2.7. Analysis of Hydraulic Jump. 27
Figure 2.8. Rectangular and Triangular Channels. 29
Figure HW2.1. Flow with Hump. 30
Figure 3.1. Rectangular Weir with Various End Contractions. 32
Figure 3.2. Triangular Weir. 33
Figure 3.3. Trapezoidal Weir. 33
Figure 3.4. Example of Spillway and Stilling Basin 34
Figure 3.5. Hydraulic Jump in Stilling Pool. 35
Figure 3.6. Examples of Plunging Pool. 39
Figure 3.7. Plunging Pool. 39
Figure 3.8. Weir Section On Top of Drop Structure 42
Figure HW3.1. Design of Check Dam. 43
Figure HW 3.2. Design of Stilling Basin. 44
Figure 4.1. Examples of Drop Structures 45
Figure 4.2. Procedure for New Slope for Permissible Velocity. 46
Figure 4.3. Continuous Drops. 47
Figure 4.4. Sloping Drop and Vertical Drop 50
Figure 4.5. Various Types of Drops 50
Figure 4.6. Various Building Materials for Drops. 51
Figure 4.7. Forces Around Drop Structure. 52

Figure 4.8. Flow over Drop Structure. 53
Figure 4.9. Layout of Drop Structure 54
Figure 4.10. Drawdown and Backup Water Surface Profiles. 56
Figure 4.11. Safety Concerns around Drop Structure. 58
Figure HW4.3. Drop Structure Design. 59
Figure 5.1. Concrete Channel. 61
Figure 5.2. Trickle Channel. 62
Figure 5.3. Superelevation Along Outer Bank in Curved Channel. 63
Figure 5.4. Weeping Hole on Concrete Bank. 64
Figure 5.5. Failures in Concrete Channels. 64
Figure 5.6. Safety and Maintenance for Concrete Channels. 65
Figure 5.7. Channel Excavated Area. 66
Figure 5.8. Transition from Straight to Curved Reaches. 67
Figure 6.1. Natural Channel. 79
Figure 6.2. Grass Linings. 81
Figure 6.3. Manning's N For Grass Linings. 83
Figure 6.4. Trickle Low-flow Channel in Grass Channel. 84
Figure 6.5. Iterative Process for Grass Channel Design. 85
Figure 6.6. Construction Stages of Grass Channel. 87
Figure 7.1. Riprap Channels. 91
Figure 7.2. Drop Protected with Riprap Rocks. 92
Figure 7.3. Riprap Toe and Bed in Cross Section. 94
Figure 7.4. Steel Anchors and Riprap Blanket. 95
Figure 7.5. Rocks used for Bank Protection. 96
Figure 7.6. Composite Riprap Lining. 97
Figure 7.7. Distribution of Shear Stress along Wetted Parameter. 98
Figure 7.8. Outer Bank Protection Along Curve Reach. 99
Figure 7.9. Composite Linings 99
Figure 7.10. Grass and Riprap Linings. 103
Figure 8.1. Multiple Land Uses on Urban Channel Floodplains. 105
Figure 8.2. Cross Section of Urban Channel. 106
Figure 8.3. Cherry Creek in Dry Time and Wet Time, Denver, Colorado. 106
Figure 8.4. Low Flow Channel within Floodway. 107
Figure 8.5. Examples of Low Flow Channel. 108
Figure 8.6. Examples of Trickle Channel. 108
Figure 8.7. Examples of Wetland Channel. 109
Figure 8.8. Low Flow Channel Capacity. 111
Figure 8.9. Hydraulic Capacity of a Composite Channel. 113
Figure 8.10. Hydraulic Capacity in Composite Channel. 114
Figure 8.11. Public Safety on and along Urban Floodplains. 117
Figure HW8.1. Floodplain Encroachment. 118

Figure HW8.2. Layout of Flood Channel. 119
Figure 9.1. Roll Waves in High Gradient Channel. 121
Figure 9.2. Cross Sectional Element in Channel Flow. 123
Figure 9.3. Flow Stability Versus Limiting Froude Number. 124
Figure 9.4. Moving Hydraulic Jump. 126
Figure 9.5. Channel Excavated Area. 129
Figure 10.1. Elements of Culvert. 131
Figure 10.2. Examples of Culverts. 132
Figure 10.3. Culvert Elements. 133
Figure 10.4. Hydraulic Parameters in a Closed Conduit. 135
Figure 10.5. Culvert Hydraulics under Inlet Control. 139
Figure 10.6. Culvert with Submerged Inlet and Outlet. 140
Figure 10.7. Design of Culvert. 141
Figure 10.8. Safety around Culvert. 143
Figure HW10.4. Design of Culvert Bridge. 144

List of Tables

Table 1.1. Recommended Manning's Roughness Coefficients. 9
Table 1.2. Example for Rating Curve Method. 13
Table 1.3. Comparison of Full Flow Capacity between Two Pipes. 13
Table 2.1. Specific Energy Curve. 19
Table 2.2. Example for Specific Force Curve. 25
Table 4.1. Design of Drop Structure. 55
Table 5.1. Manning's Roughness for Concrete Surface. 62
Table 5.2. Design Submittal Checklist for Concrete-Lined Channel. 76
Table 6.1. Retardance of Grass Linings. 82
Table 6.2. Limiting Manning's N for Grass Linings. 82
Table 6.3. Design Submittal Checklist for Grass-Lined Channel. 88
Table 7.1. Commercial Types of Riprap Rock. 93
Table 8.1. Manning's Roughness for Composite Grass Channel. 110
Table 9.1. Flow Stability Analysis for Q=1500 cfs, N=0.014, So= 2.0%. 125
Table 10.1. Allowable Headwater Depth to Culvert Diameter Ratios. 132
Table 10.2. Recommended Minimum Culvert Sizes. 133

CHAPTER 1
OPEN-CHANNEL HYDRAULICS

Open Channel Hydraulics is an applied science developed to understand the nature of open channel flows and to design infrastructures to deliver water flows. In engineering practice, a water flow is termed open channel flow wherever a free water surface exists. The free water surface profile under the atmospheric pressure represents the *hydraulic gradient line* (HGL). An open channel flow is driven by the gravity and sensitive to the hydraulic and energy gradients.

1.1 CLASSIFICATION OF CHANNELS

Water flows are carried in channels or closed conduits. Channels can be *natural* or *man-made.* A natural waterway in Figure 1.1(a) has been shaped and eroded by storm runoff flows over a geologic time. Waterways are varied from shallow overland flow, small brooks, mountain streams, rivers, to tidal estuaries. *Man-made channels* in Figure 1.1(b) are constructed for the purpose of storm water drainage, flood mitigation, irrigation, navigation, power generation etc. A *prismatic channel* is a man-made straight channel which has a constant invert slope and uniform cross section. Examples of prismatic channels include pipes, culverts, chutes, flumes, canals etc.

(a) Natural Channel	(b) Man-made Channel

Figure 1.1. Natural and Man-made Channels.

A natural stream is irregular in cross sectional width, depth, and invert slope. To estimate the flow capacity in a natural waterway, the flow cross section is divided into several vertical rectangular slots or approximated by a trapezoidal shape.

1.2. CLASSIFICATION OF CHANNEL FLOWS

Open channel flow is three dimensional in nature. On a wide floodplain, the water flow moves *laterally, longitudinally, and vertically*. In practice, the channel flow is considered one dimensional in the longitudinal direction because the velocity component along the channel alignment dictates the flow.

At a curve section, the two-dimensional flow is manifested by the superelevation that creates significant lateral movements. To study the sediment transport, the vertical flow in the gravitational direction dominates particle's settlements. For the purpose of channel design, the one-dimensional approach is sufficient for selecting the cross sectional elements. As illustrated in Figure 1.2, the channel flow is steady when the channel carries a constant discharge. The channel flow is considered *uniform* if the flow distribution remains the same at all sections, otherwise the flow is *non-uniform*.

When the acceleration is numerically insignificant in a non-uniform flow, then the flow is termed *gradually varied flow*. Otherwise, it is a *rapidly varied flow*. As illustrated in Figure 1.3, the water flow released from a reservoir into a spillway begins with a drawdown profile at the entrance and then conforms to a uniform flow farther downstream. At the foot of the spillway, a hydraulic jump is triggered, according to the tailwater. The drawdown profile represents a gradually varied flow (GVF) and a hydraulic jump is a rapidly varied flow (RVF).

Using a sluice gate to control the flow release, the discharge in the channel is adjusted from time to time. When the discharge in a channel changes with respect to time, the flow is classified as *unsteady*, otherwise the flow is steady. The changes in flow velocity are termed acceleration. A *local acceleration* is the change in flow velocity at a specified cross section over a time interval while a *convective acceleration* is the change in flow velocity over a distance at a specified time. The former may be caused by the variations in flow released from the upstream reservoir into the channel, and the latter may be caused by the expansions or contractions in the channel width along the floodplain.

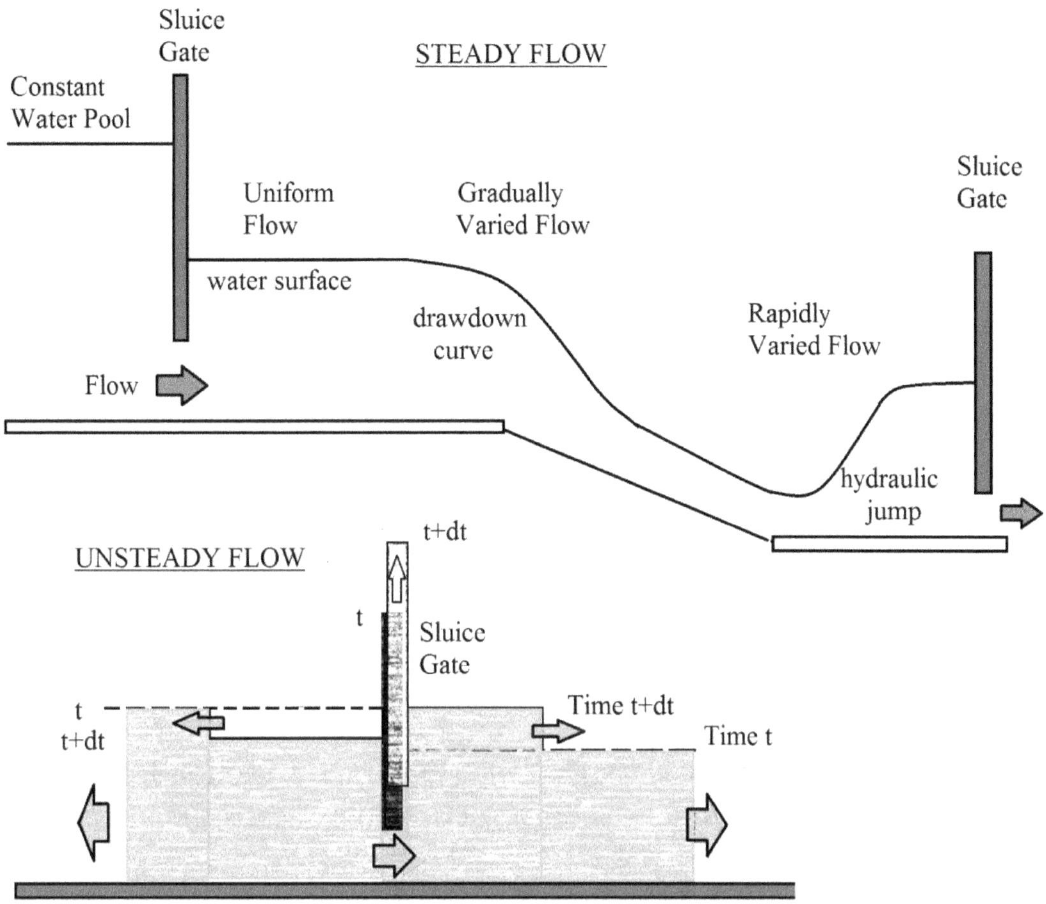

Figure 1.2. Classifications of Channel Flows.

1.3. SLOPES IN CHANNEL FLOW

Channel flow is driven by the energy difference or gradient. The Bernoulli's sum at a section in a channel flow is defined as:

$$E = Y + Z + Hv = Y + Z + \frac{U^2}{2g} \qquad (1.1)$$

Where E= Bernoulli's sum in [L], Y= flow depth in [L], Z= channel invert elevation in [L], *Hv*= velocity head in [L], U= cross sectional velocity in [L/T], g= gravitational acceleration in [L/T^2]. As illustrated in Figure 1.3, between two adjacent sections, the energy slope is computed as:

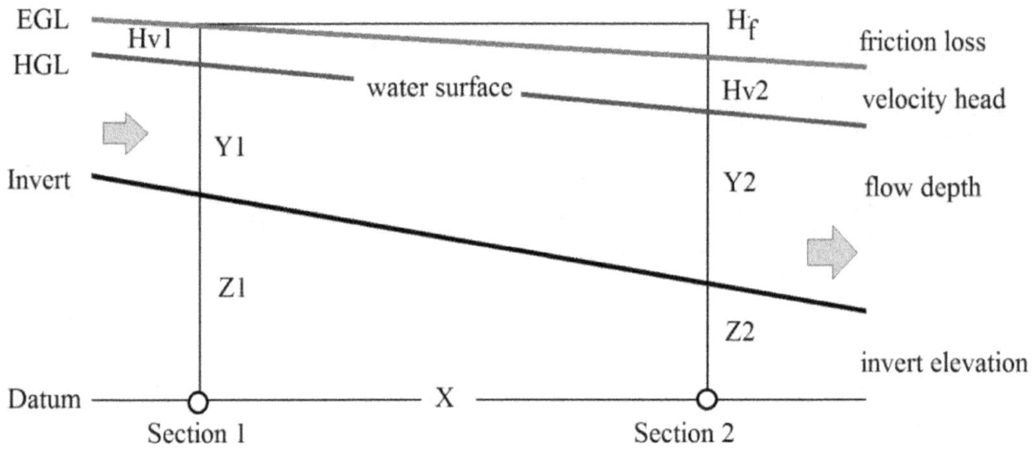

Figure 1.3. EGL and HGL in Channel Flow.

$$S_e = \frac{H_f}{X} = \frac{E_2 - E_1}{X} \quad (1.2)$$

Where S_e= energy slope on *Energy Grade Line* (EGL), H_f= friction loss in [L], and X= distance in [L] between two sections. The subscription 1 means the variable associated with Section 1, etc.

In a channel flow, the water surface is the *Hydraulic Grade Line* (HGL). The slope of HGL is calculated as:

$$S_W = \frac{(Y_2 + Z_2) - (Y_1 + Z_1)}{X} \quad (1.3)$$

Where S_w= slope of water surface in [L/L]. The channel invert slope is calculated as:

$$S_o = \frac{Z_2 - Z_1}{X} \quad (1.4)$$

Under the assumption that the acceleration in the channel flow is negligible, the flow is in equilibrium and termed uniform flow or normal flow. A uniform flow is characterized with:

$$S_e = S_w = S_o \quad (1.5)$$

Design of a channel begins with the known ground slope. Channel runs from a high point into a downstream water body. As illustrated in Figure 1.4, between two points, it is a good choice to set the channel invert slope to be the ground slope. The existing irregular cross sections serve as guidance to select regular cross sections for channel improvements. Often in urban areas, banks and levees are used to reduce the width of the floodplain. The floodplain encroachments need to observe the floodplain regulations and design criteria.

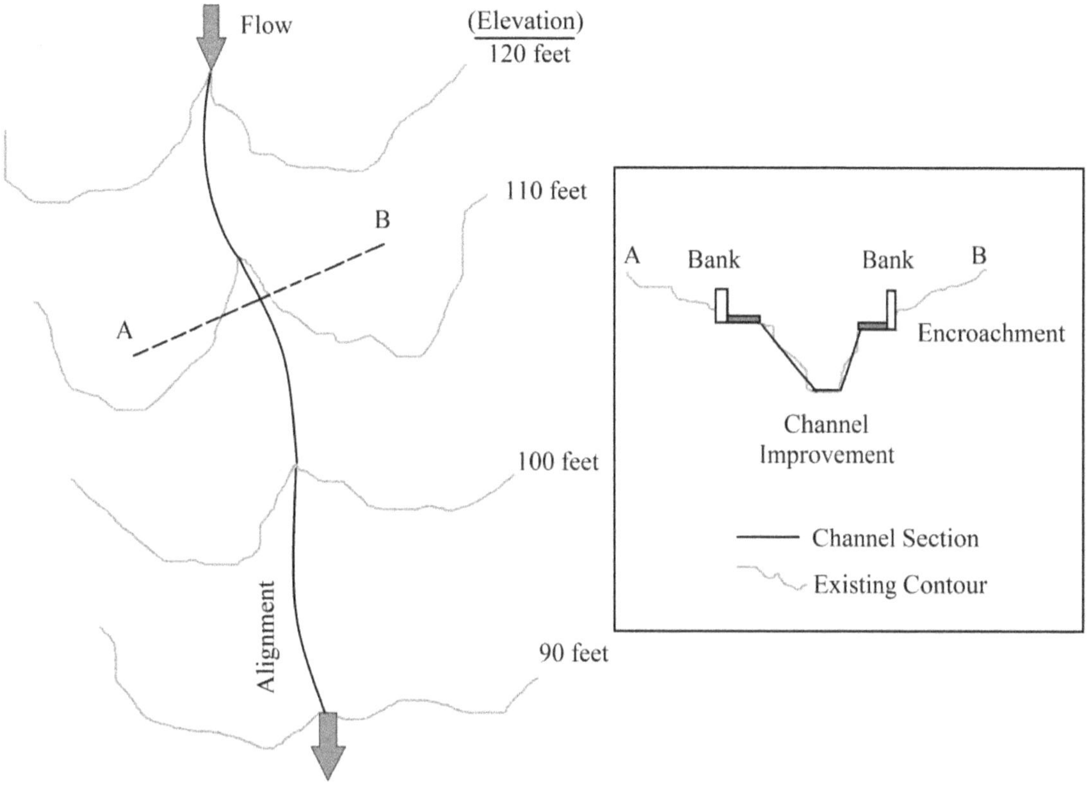

Figure 1.4. Channel Alignment Between Points.

Many empirical formulas were developed to predict cross sectional average flow velocity in a channel. In general, they were derived under the assumption of one-dimensional uniform flow. Empirical formulas developed for channel flows are derived for depth-integrated parameters for channel flow. Therefore, a channel flow is dictated by its cross sectional elements.

1.4 CROSS SECTIONAL ELEMENTS

Most man-made channels are trapezoidal in shapes. Both rectangular and triangular channels are special cases. The flow condition in a channel is closely related to the channel cross sectional geometry. Elements in a trapezoidal cross section are illustrated in Figure 1.5, including:

(1) *flow depth*, Y, which is the vertical distance from the water surface to the lowest point on the cross,

(2) *bottom width, B,*

(3) *flow area,* A, which is the cross sectional area that conveys the flow,

(4) *side slope include:* Z_1 on left side *and* Z_2 on right side,

(5) *wetted perimeter, P, which* is the length of the wetted surface,

(6) *top width, T, which is* the width on water surface,

(7) water surface elevation (or water stage), Ws,

(8) *channel bottom elevation, Z.*

(9) freeboard, F, which is the vertical distance above the water surface to bank top.

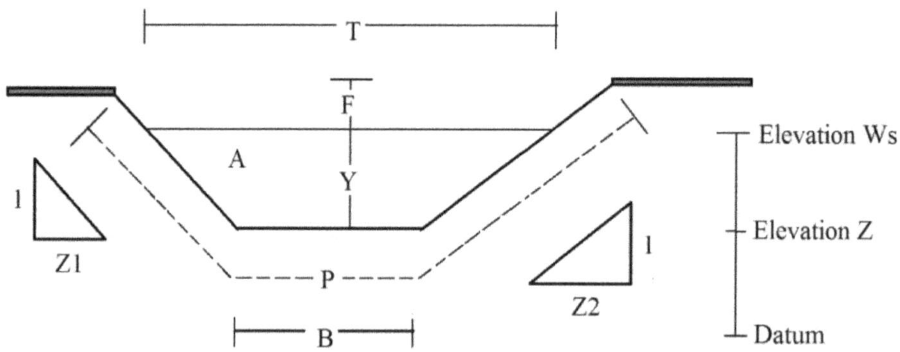

Figure 1.5. Cross Sectional Elements in Trapezoidal Channel.

For a trapezoidal channel, the flow area and wetted perimeter are calculated as:

$$T = B + (Z_1 + Z_2)Y \quad (1.6)$$

$$A = \frac{1}{2}[B + T]Y = [B + \frac{1}{2}(Z_1 + Z_2)]Y \quad (1.7)$$

$$P = B + [\sqrt{1 + Z_1^2} + \sqrt{1 + Z_2^2}]Y \quad (1.8)$$

When the values of Z_1 and Z_2 are reduced to zero, we have a *rectangular channel*. Similarly, a trapezoidal channel is reduced to a *triangular channel* when the bottom width, B, vanishes.

1.5. EMPIRICAL FORMULA

A *normal flow* is a special case that can be developed in a long and straight prismatic channel with a constant discharge. Normal flow is also termed uniform flow. In a uniform flow, its energy slope, S_e, is parallel to its bottom slope, S_o. In 1889, Robert Manning presented a formula to predict the normal flow condition in a channel. Manning's equation states:

$$U_N = \frac{K_N}{N} R_N^{\frac{2}{3}} \sqrt{S_o} \tag{1.9}$$

$$Q = U_N A_N \tag{1.10}$$

in which K_N= 1.486 for feet-second units or 1.0 for meter-second units, N = Manning's roughness, R_N= hydraulic radius in [L], U_N= normal flow velocity in [L/T], and A_N= flow area in [L^2|, and Q= discharge in [L^3/T]. The subscript, n, represents variables associated with normal flow condition. A channel flow is sensitive to the surface friction that directly applies to the wetted surface. *Hydraulic radius*, R, was developed as the ratio of flow area to wetted perimeter as:

$$R_N = \frac{A_N}{P_N} \tag{1.11}$$

Where P_N= wetted perimeter in [L]. The discharge carried in a channel is directly related to its water depth. An increase in the top width in the channel means an increase in the discharge. *Hydraulic Depth*, H_d, is defined as the ratio of flow area to top width as:

$$H_d = \frac{A_N}{T_N} \tag{1.12}$$

Where T_N= top width of water surface in [L]. As implied in the definitions of cross sectional elements, the flow capacity in a channel is determined by the flow cross sectional area, roughness on the wetted perimeter, and energy gradient. Substituting Eq's 1.10 and 1.11 into Eq 1.9 yields:

$$Q = \frac{K_N}{N} A_N^{\frac{5}{3}} P_N^{\frac{-2}{3}} \sqrt{S_o} \tag{1.13}$$

There are seven variables involved in Eq. 1.13. They are discharge, channel slope, flow depth, bottom width, right side slope, left side slope, and Manning's roughness. After selecting any six variables, the only unknown is determined by Eq. 1.13.

In practice, open channel flows are classified by two dimensionless parameters: Reynolds number and Froude number. *Reynolds number* is the ratio of flow inertial force to viscous force and defined as:

$$R_e = \frac{U_N R_N}{\upsilon} \tag{1.14}$$

in which R_e = Reynolds number, and ν = water kinematic viscosity in $[L^2/T]$. Reynolds number serves as an indicator to differentiate *laminar* flows from *turbulent* flows. A laminar open channel flow shall have a Reynolds number less than 500. In practice, it is rare to find a laminar open channel flow in nature.

Froude number is the ratio of flow inertial force to weight of water. It also represents the ratio of flow velocity to wave celerity. Froude number is defined as:

$$F_r = \frac{U_N}{\sqrt{gH_d}} \tag{1.15}$$

Many characteristics of open-channel flows are associated with Froude number. For instance, when Fr<1, the flow is termed *subcritical flow. When* Fr>1, the flow is called *supercritical flow. A critical flow* has a Froude number equal to unity. Froude number is often used in the similarity rules for modeling studies. Details about these three regimes of open-channel flows will be discussed later.

Example 1.1: A concrete box conduit of 10-ft wide and 7 ft height, as illustrated in Figure 1.6, carries a flow of 420 cfs. Determine the slopes associated with the flow.

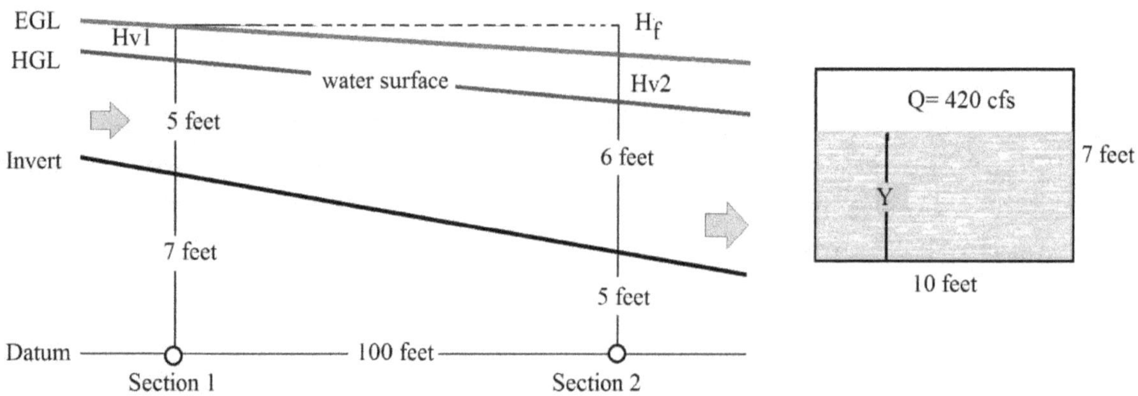

Figure 1.6. Water Flow in Concrete Box.

At Section 1, the bottom elevation of the box conduit is at 7 feet and the flow depth is 5 feet. At the Section 2, the bottom elevation of the box conduit is at 5 feet and the flow depth is 6 feet. With a distance of 100 ft between these two sections, the bottom slope, water surface slope, and flow energy slope are calculated as:

$$S_o = \frac{5.0 - 7.0}{100.0} = -0.02$$

$$S_w = \frac{(6.0 + 5.0) - (5.0 + 7.0)}{100.0} = -0.01$$

$$420 = U_1(5 \times 10) = U_2(6 \times 10) \quad U_1 = 8.4\ fps \text{ and } U_2 = 7.0\ fps$$

$$S_e = \frac{(6.0 + 5.0 + \frac{7.0^2}{2 \times 32.2}) - (5.0 + 7.0 + \frac{8.4^2}{2 \times 32.2})}{100.0} = -0.013$$

1.6. ROUGHNESS COEFFICENT

Although Manning's formula is widely accepted, caution must be taken when estimating the roughness coefficient. Equation 1.10 indicates that the channel capacity is linearly and reversely varied with respect to Manning's N. A change from N=0.04 to N=0.02 means the flow rate is doubled. In practice, Manning's N is mainly selected based on the lining materials. For design, Table 1.1 is recommended for the selection of Manning's N.

Table 1.1. Recommended Manning's Roughness Coefficients.

Channel Lining Description	*Manning's Roughness N*
Cement	0.011- 0.013
Concrete	0.012-0.015
Wood	0.014
Brick	0.015
Asphalt	0.016
Masonry	0.017
Earth	0.022
Vegetal Lining	0.030-0.035
Rock Riprap	0.035-0.045
Natural Floodplain	0.045-0.050
Natural Channel	0.030-0.035
Flood Plain	0.050-0.080
Bushes	0.070
Trees	0.090

On a wide floodplain, Manning's N is varied from the left bank to the right bank or from the bottom to the water surface. For the purpose of design, a single value for the entire cross section is acceptable for Manning's N.

1.7. NORMAL FLOW

Figure 1.7(a) presents a case of prismatic channel in which a uniform flow can be developed. On the contrary, the curve channel in Figure 1.7(b) will produce non-uniform flows due to the superelevation effects along the outer bank.

(a) Straight Prismatic Channel	(b) Curve Channel

Figure 1.7. Uniform and Non-uniform Flows.

A normal flow is characterized with a constant flow depth and constant velocity everywhere for the given discharge. It implies that the external forces acting on the uniform flow are in equilibrium. In practice, hardly can we find a uniform and steady flow in a natural waterway because the flow rates and cross sections change with respect to time and to space as well. Despite this fact, the design procedure begins with the normal flow.

Example 1.2: A trapezoidal channel is specified in Figure 1.8. The flow depth is 5 feet. Determine the flow area, A, top width, T, wetted perimeter, P, hydraulic radius, R, and hydraulic depth, D.

$$T = B + (Z_1 + Z_2)Y = 10 + (3+4)\times 5 = 45 \text{ ft}$$

$$A=\frac{B+T}{2}Y=\frac{10+45}{2}\times 5=137.5 \quad \text{sq ft}$$

$$P=B+(\sqrt{1+Z_1^2}+\sqrt{1+Z_2^2})Y=10+(\sqrt{1+3^2}+\sqrt{1+4^2})\times 5=46.43 \quad \text{ft}$$

$$R=\frac{A}{P}=2.96\,\text{ft}$$

$$H_d=\frac{A}{T}=\frac{137.5}{45}=3.06\,\text{ft}$$

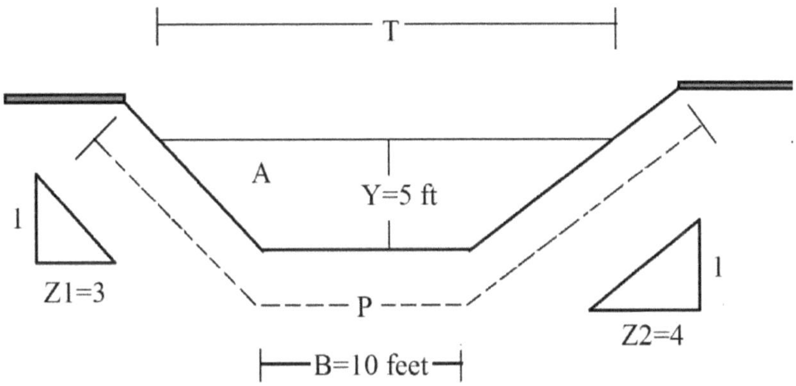

Figure 1.8. Flow Parameters for Example.

Example 1.2 reveals that the flow depth is neither the hydraulic depth nor the hydraulic radius. For a wide channel which is defined as B>10Y, numerically, we may have $Y \approx R \approx H_d$.

***Example 1.3*:** A trapezoidal channel carries a discharge of 600 cfs. The channel has a slope of 0.01, roughness of 0.035, bottom width of 5 ft, and side slope of *1V:ZH, Z=4.0*. Determine the normal depth.

***Solution*:** Substituting design parameters into Eq's 1.6 , 1.7, and 1.8 yields:

$$T_n=5.0+(4.0+4.0)Y_n$$

$$A_n=\frac{5.0+T}{2}Y_n=5.0Y_n+4.0Y_n^2$$

$$P_n=5.0+2Y_n\sqrt{1+4.0^2}=5.0+8.25Y_n$$

The subscript of "n" represents the variables associated with the normal flow. Substituting the above flow variables into Eq 1.13 yields:

$$Q = \frac{1.486}{0.035}[5.0Y_n + 4.0Y_n^2]^{5/3}(5.0 + 8.25Y_n)^{-2/3}\sqrt{0.01} = 600\,cfs$$

The above equation is nonlinear with respect to flow depth. The solution can be found by trial and error for Q=600 cfs. In practice, it is convenient to utilize the *rating curve method* to solve Manning's equation. A rating curve represents the relationship between flow depths and flow rates using Eq 1.13. A rating curve is derived for the specified cross section, invert slope and Manning's roughness.

Table 1.2 presents the solution for this case. The flow depth varies within the selected range. The rating curve is plotted in Figure 1.9. The solution for this case can be determined by interpolation. So we have $Y_n = 3.98\ ft$, $A_n = 83.3\ ft^2$, $P_n = 37.8\ ft$, and $U_n = 7.2\ fps$. The corresponding hydraulic depth and Froude number are:

$$F_r = \frac{7.20}{\sqrt{32.2 \times 3.98}} = 0.84$$

It is a subcritical flow because $F_r = 0.84 < 1.0$.

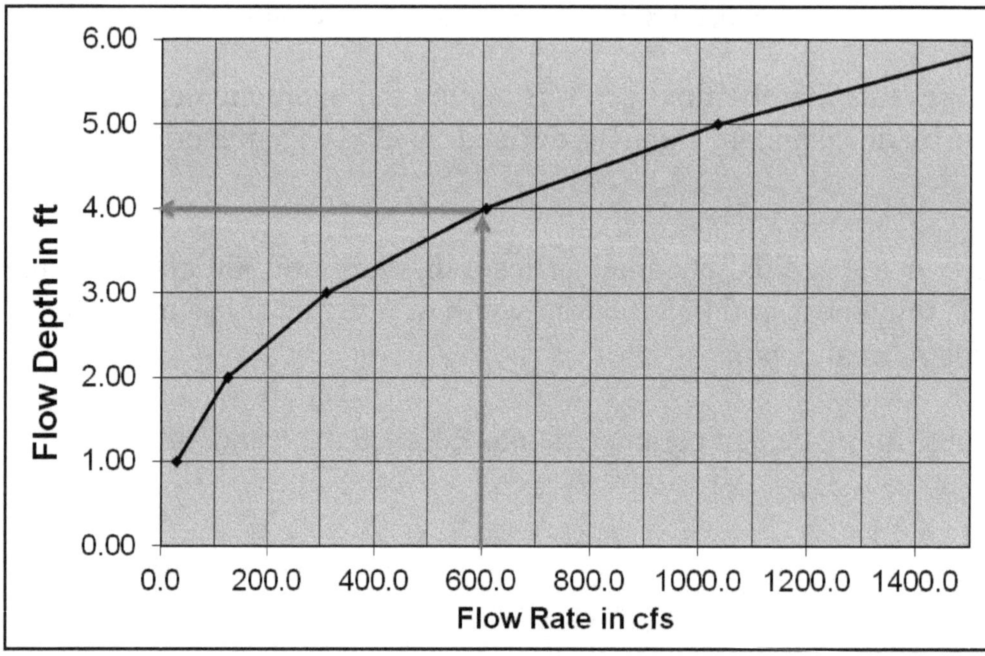

Figure 1.9. Example Rating Curve.

Table 1.2. Example for Rating Curve Method.

Channel Bottom Width	B=	5.00	feet
Left Side Slope	Z1=	4.00	ft/ft
Right Side Slope	Z2=	4.00	ft/ft
Manning's n	N=	0.035	
Channel Bottom Slope	So=	0.0100	ft/ft

Flow Depth Y ft	Top Width Top ft	Flow Area A sq ft	Wetted P-meter P ft	Hydraulic Radius R ft	Flow Velocity U fps	Flow Rate Q cfs	Froude Number Fr
1.00	13.00	9.00	13.25	0.68	3.28	29.5	0.69
2.00	21.00	26.00	21.49	1.21	4.82	125.4	0.76
3.00	29.00	51.00	29.74	1.71	6.09	310.8	0.81
4.00	37.00	84.00	37.98	2.21	7.23	607.0	0.85
5.00	45.00	125.00	46.23	2.70	8.27	1033.4	0.87

Example 1.4: An old corrugated metal pipe of 36 inches in diameter is laid on a slope of 2%. After years of service, this circular pipe needs to be replaced. The Manning's N for this old pipe is 0.035. Verify that a 27-inch concrete pipe can provide similar flowing full capacity to the old 36-inch pipe.

Solution: For a circular pipe, the flowing full area and wetted parameters are calculated as:

$$A = \frac{\pi D^2}{4} \tag{1.16}$$

$$P = \pi D \tag{1.17}$$

$$R = \frac{D}{4} \tag{1.18}$$

Substituting Eq's 1.16 and 1.17 into Eq. 1.13 yields Table 1.3 for both pipes as:

Table 1.3. Comparison of Full Flow Capacity between Two Pipes.

Diameter feet	Pipe Line Slope ft/ft	Full-flow Area sq feet	Wetted Perimeter feet	Hydraulic Radius feet	Manning's N	Full-flow Discharge cfs
3.00	0.02	7.07	9.42	0.75	0.035	35.13
2.25	0.02	3.98	7.07	0.56	0.014	40.78

The proposed new pipe of 27-inch in diameter provides more flow capacity than the existing.

1.8. CLOSING

Channel design begins with the normal flow. A normal flow is defined by Manning's formula under the assumption that S_e=S_o. Solutions for a normal flow are achieved using the rating curve method which is a one-to-one increasing function between flow rates and flow depths. The normal flow approach provides the first estimate of channel dimension and capacity. As always, refinements are required when the channel alignment is curved or/and channel width is changed.

In the field, a stage-flow stream gage as illustrated in Figure 1.10 and observed in Figure 1.11, is installed to measure and to record the variations of flows in the channel. Often a stream gage only reports the instantaneous water depth. The corresponding flow rate is then read off from the rating curve.

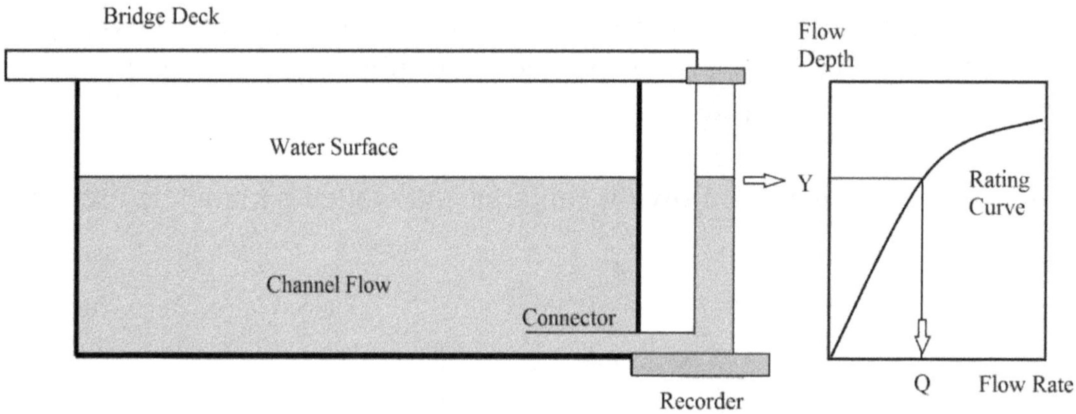

Figure 1.10. Rating Curve at Stream Gage.

A stream gage, in Figure 1.11b, shall be installed at a stable cross section. If the cross section is subject to severe erosion or backwater effects, the rating curve will not accurately represent the true flow condition. Calibrations are required to maintain the integrity of a stream gage. In addition to stream gage, we also install weirs and orifices to record the water depths. A time series of water depths can be converted into its flow hydrographs using the rating curve pre-developed for the measurement instrument.

(a) Stage-Flow Recorder | (b)Stage Station on Bank Wall

Figure 1.11. Stage-Flow Measurement.

1.9. HOMEWORK

Q1.1: Repeat Example 1.1 with the flow in a trapezoidal channel: B= 5 feet, N=0.045 and $Z_1=Z_2=2$. Determine the energy slope and water surface slope.

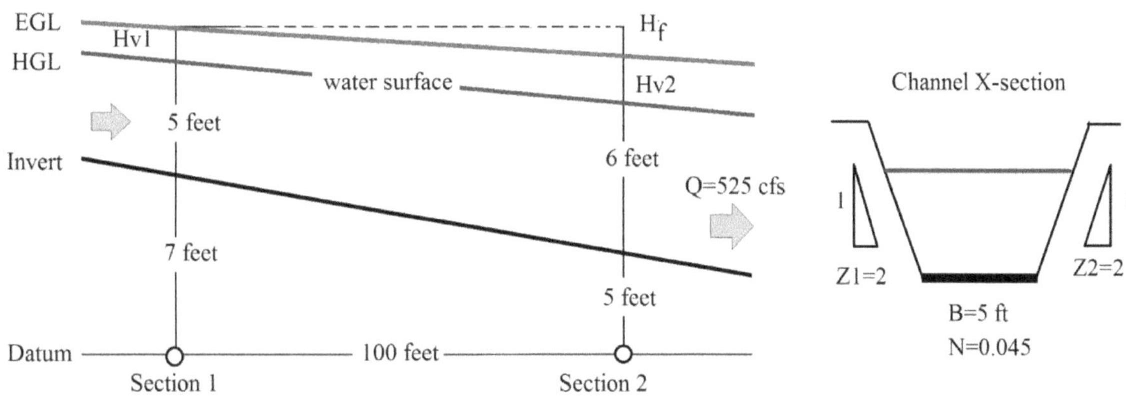

Figure HW1.1. Slopes on HGL and EGL.

Q1.2: A concrete trapezoidal channel is laid on a slope of 0.01. The bottom width is 10 ft and the side slopes are $Z_1=Z_2=1.0$. Determine the normal flow rating curve for this channel.

CHAPTER 2
HYDRAULIC DROP AND JUMP

As illustrated in Figure 2.1, when the water flow is released from a spillway into a stilling pool, the incoming supercritical flow is sustained on the high gradient while the outgoing subcritical flow is formed in the pool. In between, a *hydraulic jump* is triggered across the discontinuous water surface from the shallow to the deep depth. Similarly, when the water flow crosses a floor drop built across the channel bottom, a *hydraulic drop* is induced from the deep to the shallow depth. Hydraulic jump is the transitional process when the flow regime changes from a supercritical to a subcritical flow while a hydraulic drop is another transitional process from a subcritical to supercritical flow. The former is an abrupt change on the channel floor involving tremendous energy dissipation while the latter is a smooth transition.

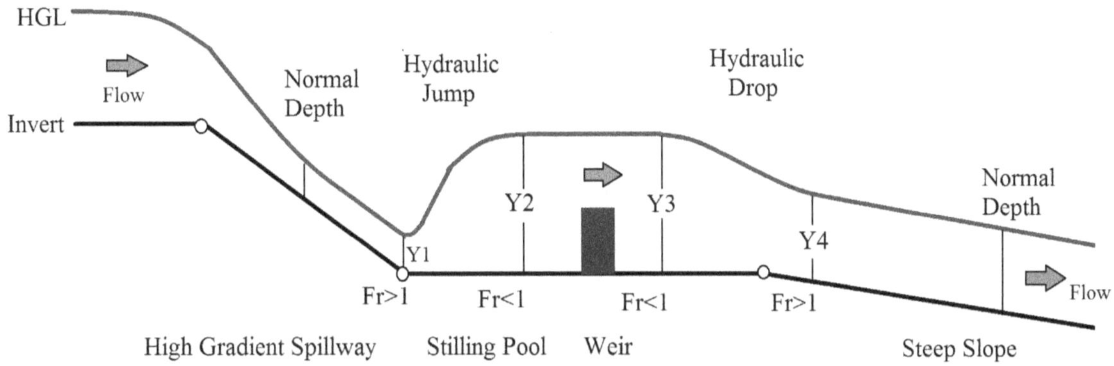

Figure 2.1. Hydraulic Jump and Drop.

Both hydraulic jump and drop occur within a flat and short pool. As a result, the change in channel's invert elevation is negligible. In this Chapter, the concepts of specific energy and force are developed to balance the Bernoulli's sum between flow depth and dynamic head. Applications of specific energy and force are often directly related to the design of stilling pool for energy dissipation.

2.1. SPECIFIC ENERGY

The basic principles for open channel flow were derived using a depth-integration method in which the flow properties are directly correlated to the cross sectional average velocity. The Bernoulli's sum at a cross section is defined as:

$$E = Y + Z + \frac{\alpha U^2}{2g} \tag{2.1}$$

$$W_S = Y + Z \tag{2.2}$$

in which E = Bernoulli sum in [L], Z = invert elevation in [L], Y= water depth in [L] from water surface to the lowest point in the cross section, U= cross sectional flow velocity in [L/T], α = energy correction factor, and W_S = hydraulic head or water surface elevation in [L]. The Bernoulli's sum at a cross section consists of *hydraulic head* and *dynamic head.* The hydraulic head for all points on the cross section is a constant, but the point flow velocity varies with respect to the water depth. As a result, the hydraulic head is represented by the sum of (Y+Z) for all points on a cross section, but the dynamic head needs a numerical adjustment, α in Eq 2.1 when using the cross sectional average velocity head. In practice, the value of α varies from 1.0 to 1.2. For simplicity, α = 1 is acceptable for practice.

At a given cross section, the invert elevation is specified. The specific energy, E_S, is defined as the sum of flow depth and velocity head as:

$$E_S = Y + \frac{\alpha U^2}{2g} \approx Y + \frac{U^2}{2g} \tag{2.3}$$

$$U = \frac{Q}{A} \tag{2.4}$$

In which Q= flow rate in [L^3/T] channel. Substituting Eq. 2.4 into Eq 2.3 yields:

$$E_s = Y + \frac{\alpha Q^2}{2gA^2} \tag{2.5}$$

Equation 2.5 represents the *specific energy curve* that is the relationship between flow depth and velocity head for a given discharge in a specific cross section. Although Eq 2.5 is a cubic equation, only the two roots of real numbers are meaningful for engineering practices. It implies that for a given specific energy, the water flow can be either a high stage with a low velocity head or a low stage with a high velocity head.

Example 2.1: A water flow of 450 cfs is delivered in a trapezoidal channel with a bottom width, B = 10 feet, and side slopes, $Z_1=Z_2=1.0$. Construct the specific energy curve for water depths ranging from 1.0 to 15.0 feet.

Solution: Direct solutions for Eq 2.5 are listed in Table 2.1. As the flow depth increases, the top width and flow area increases while the flow velocity and dynamic head (or kinetic energy) decreases. As shown in Figure 2.2, the specific energy, E_S, reaches its minimum when Fr=1 at Y=3.52 feet.

Table 2.1. Specific Energy Curve.

Flow Depth Y ft	Top Width T ft	Flow Area A sq ft	Flow Velocity u fps	Kinetic Energy u^2/2g ft	Sp. Energy Es ft	Froude Numer Fr
1.00	12.00	11.00	40.91	25.99	26.99	7.53
3.00	16.00	39.00	11.54	2.07	5.07	1.30
5.00	20.00	75.00	6.00	0.56	5.56	0.55
7.00	24.00	119.00	3.78	0.22	7.22	0.30
9.00	28.00	171.00	2.63	0.11	9.11	0.19
11.00	32.00	231.00	1.95	0.06	11.06	0.13
13.00	36.00	299.00	1.51	0.04	13.04	0.09
15.00	40.00	375.00	1.20	0.02	15.02	0.07

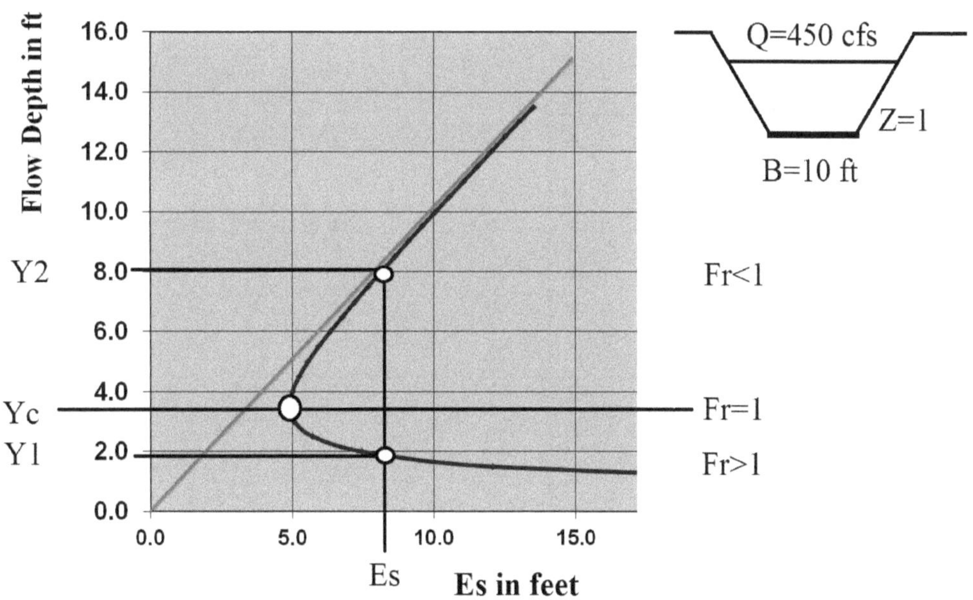

Figure 2.2. Specific Curve.

Figure 2.2 presents the specific energy curve for the given discharge in the specified channel section. The specific energy curve consists of three flow regimes identified by Froude numbers. They are *subcritical, critical* and *supercritical* flow regimes. As illustrated in Figure 2.2, the specified energy curve consists of two limbs because there are two roots, (Y_1 and Y_2) for a given specific energy, E_S. The high stage, Y_1, has a Froude number less than unity and the low stage, Y_2, has a Froude number greater than unity. The depth of Y_1 is the *alternate depth* of Y_2, and vice versa. The high stage limb represents the subcritical flow regime while the low stage limb represents the supercritical flow regime. The interception of these two limbs is the critical flow. It is noted that the specific energy at the critical flow is a minimum. Application of specific energy curve is best explained by the subcritical flow overtopping a submerged dam. The energy loss is for such a mild drop is negligible. As a result, both Y_1 and Y_2 in Figure 2.3 share the same specific energy as:

$$E_1 = E_2 + \Delta E \approx E_2 \tag{2.6}$$

For the design flow, the critical depth occurs on the top of the dam.

$$E_1 = E_2 = E_c + H_m \tag{2.7}$$

in which E_1= specific energy in [L] at Section 1, E_2= specific energy in [L] at Section 2, ΔE= energy loss in [L], E_c = specific energy in [L] for critical flow, and H_m= height in [L] of submerged dam.

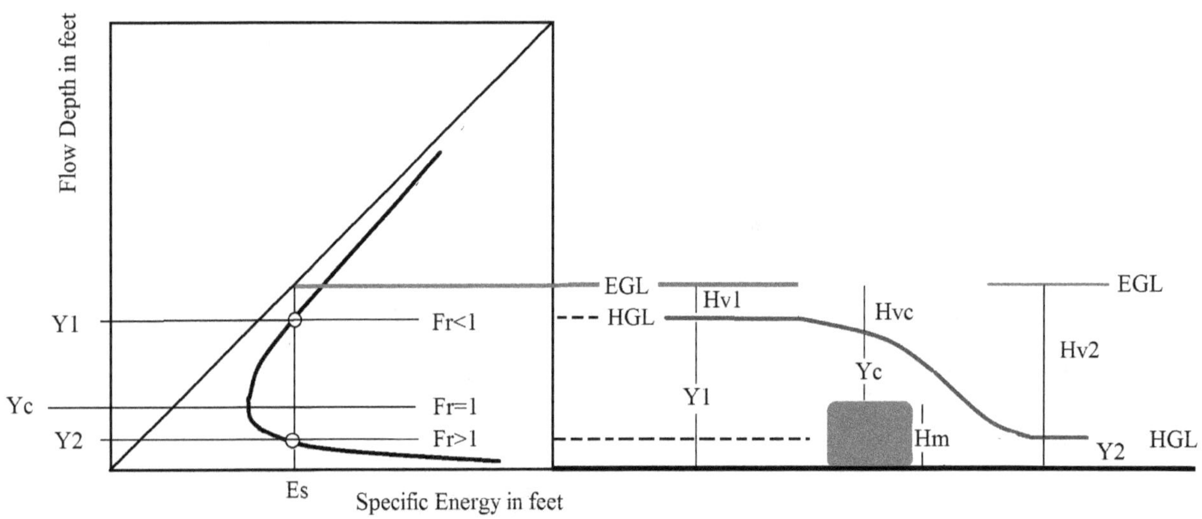

Figure 2.3. Application of Specific Energy Curve.

2.1.1. Critical Flow on Specific Energy Curve

Channel flows are classified by Froude number that is defined as the ratio of flow velocity to wave celerity as:

$$F_r = \frac{U}{\sqrt{gH_d}} = \sqrt{\frac{\alpha U^2 T}{gA}} = \sqrt{\frac{\alpha Q^2 T}{gA^3}} \qquad (2.8)$$

As illustrated in Figure 2.3, the specific energy becomes a minimum at the critical depth. To prove it, taking the first derivative of Eq 2.3 with respect to depth Y yields:

$$\frac{dE_s}{dY} = 1 - \frac{\alpha U^2 T}{gA} = 1 - \frac{\alpha U^2}{gH_d} = 1 - F_r^2 = 0 \qquad (2.9)$$

Equation 2.9 satisfies that Fr=1 at the critical depth. Aided with Eq 2.9, the critical hydraulic depth is derived as:

$$\frac{(H_d)_c}{2} = \frac{\alpha U_c^2}{2g} \qquad (2.10)$$

in which $(H_d)_c$ = critical hydraulic depth and U_c = critical flow velocity. For a rectangular channel, $(H_d)_c = Y_c$ which is the critical depth·

2.1.2. Special Case for Specific Energy

With Z=1, a trapezoidal channel is reduced to a rectangular. When Z=0, then H_d=Y. Equation 2.10 is further reduced to:

$$\frac{Y_c}{2} = \frac{\alpha U_c^2}{2g} \quad \textit{for rectangular channel} \qquad (2.11)$$

Substituting the continuity principle into Eq 2.11 yields:

$$Y_c = (\frac{\alpha q^2}{g})^{1/3} \approx (\frac{q^2}{g})^{1/3} \quad \textit{for rectangular channel} \qquad (2.12)$$

$$q = \frac{Q}{B} \quad \textit{for rectangular channel} \qquad (2.13)$$

in which q = discharge per unit width in [L^2/T]. Aided with Eq 2.11, the critical depth in a rectangular channel is two thirds of the specific energy.

$$E_c = Y_c + \frac{\alpha U_c^2}{2g} = Y_c + \frac{Y_c}{2} = \frac{3Y_c}{2} \quad \textit{for rectangular channel} \tag{2.14}$$

Example 2.2: Find the critical flow depth for the channel flow in Example 2.1.

Solution: The critical depth is solved by setting the flow Froude number equal to one as:

$T_c = B + 2ZY_c = 10.0 + 2.0\,Y_c$

$A_c = (B+ZY_c)Y_c = (10.0 + 1.0\,Y_c)\,Y_c$

$$F_r^2 = \frac{Q^2 T_c}{gA_c^3} = \frac{450^2(10.0+2.0Y_c)}{32.2[(10.0+1.0\times Y_c)Y_c]^3} = 1$$

By trial and error, we have $Y_c = 3.52$ *ft*, $A_c = 47.5$ *ft*^{2}, *and* $U_c = 9.47$ *fps*.

Example 2.3: The flow in Example 2.1 is carried on a slope of 0.5% with Manning's N=0.045. Referring to Figure 2.3, the incoming flow is set to be the normal flow, $Y_1=Y_n$. (1) Determine the alternate depth, Y_2, downstream of the submerged dam. (2) What is the height of the dam?

Solution: The normal flow condition is solved as:

Flow Depth Yn ft	Flow Area An sq ft	Wetted P-meter Pn ft	Hy-Radius Rn ft	Flow Velocity Un fps	Flow rate Q cfs	Froude Numer Fr
5.53	85.8	25.6	3.35	5.25	450.0	0.46

Referring to Figure 2.3, set $Y_1=Y_n$ to determine E_s=5.95 feet and incoming flow Fr<0.46. It is confirmed that the incoming flow is subcritical. After the drop the flow becomes supercritical or $Y_2<Y_C$. The trial and error procedure may begin with a value< 3.52 feet for Y_2. For this case, it is found that Y_2=2.40 feet and Fr=1.88 can produce the same specific energy as E_s=5.95 feet.

Station	Flow Depth Y ft	Flow Area A sq ft	Top Width T ft	Flow Velocity u fps	Kinetic Energy u^2/2g ft	Sp. Energy Es ft	Froude Numer Fr
Y=Y1=Yn	5.53	85.8	21.05	5.25	0.43	5.95	0.46
Y=Y2	2.40	29.8	14.80	15.12	3.55	5.95	1.88
Y=Yc	3.52	47.5	17.03	9.47	1.39	4.91	1.00

The balance of flow energy between Section 1 and the critical flow section is:

$E_{s1}=E_{sc}+H_m$

$5.96=4.91+H_m$ So, $H_m= 1.05$ *ft.* which is the high of the submerged dam.

2.2. SPECIFIC FORCE

The momentum force associated with a channel flow consists of static force and dynamic force. The hydrostatic force results from the pressure (weight) of water body while the dynamic force is force required to bring the flow to a stop.

$$F_{HS} = \gamma Y_o A \tag{2.15}$$

$$F_{HD} = \beta \rho Q U \tag{2.16}$$

$$F_S = \gamma Y_o A \beta + \rho Q U \tag{2.17}$$

in which F_{HS}= hydro static force, F_{HD}= dynamic force, F_S= specific force, γ = specific weight of water, Y_o = the vertical distance in [L] to centroid of flow area measured from the surface, A= flow area in [L^2], *i= i-th* area element, ρ = density of water in [M/L^3], β = momentum correction factor varied in a range of 1.0 to 1.1. All forces and weight are expressed in pounds or newtons. The value of β was developed to count for the non-uniformity of turbulent velocity distribution. In practice, let β be 1.0. The flow area can be divided into regular area elements. The location of the centroid of the flow area is determined by the area-weighted method as:

$$Y_o = \frac{\sum Y_{oi} A_i}{A} \tag{2.18}$$

Example 2.4: The trapezoidal channel in Example 2.1 carries a discharge of 450 cfs with Y=4.5 ft, A=65.25 sq feet, and U=6.90 ft/second. The flow area is divided into two triangles and one rectangle in Figure 2.4. Determine the momentum force in pounds.

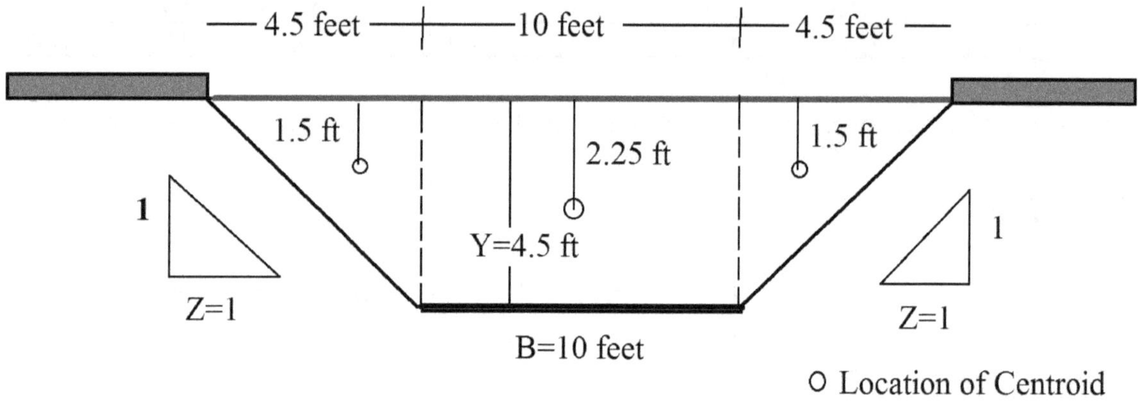

Figure 2.4. Calculating Specific Force.

Solution:

Apply the area-weighted method to the three area elements to determine the centroid for the flow area as:

$$Y_o = \frac{\sum Y_{oi} A_i}{A} = \frac{1.5\times\frac{1}{2}(4.5\times4.5)+2.25\times(4.5\times10.0)+1.5\times\frac{1}{2}(4.5\times4.5)}{65.25} = 2.01\ ft$$

The specific force carried by the flow is:

$$F_S = \gamma Y_o A + \rho\beta Q U = (62.4\times2.01\times65.25+1.92\times1.0\times450\times6.90)/1000 = 14.22\,klb$$

Repeat the above procedure for flow depths ranging from 1.0 to 6.5 feet. Table 2.2 is the specific force curve for this case. It is noted that the critical depth is approximately 3.50 feet.

Similar to the specific energy curve, the specific force curve is developed for a given discharge in a specified channel cross section. A specific curve consists of three flow regimes: subcritical flow, critical flow, and supercritical flow. As illustrated in Figure 2.5, the upper limb represents the subcritical flow regime and the lower limb represents the supercritical regime. Both limbs intercept each other at the critical depth. Application of a specific force curve is best explained by the water flow released from a steep spillway that results in a hydraulic jump in the stilling basin. Currently, the huge dissipation through a hydraulic jump is not fully understood. Therefore a hydraulic jump

Table 2.2. Example for Specific Force Curve.

Flow Depth Y ft	Top Width T ft	Flow Area A sq ft	Flow Velocity u fps	Central Depth Yo ft	Sp. Force Fs Klbs	Froude Numer Fr	SP Energy Es ft
1.00	12.00	11.00	40.91	0.48	36.05	7.53	26.99
1.50	13.00	17.25	26.09	0.72	23.55	3.99	12.07
2.00	14.00	24.00	18.75	0.94	17.78	2.52	7.46
2.50	15.00	31.25	14.40	1.17	14.84	1.76	5.72
3.00	16.00	39.00	11.54	1.38	13.44	1.30	5.07
3.50	17.00	47.25	9.52	1.60	13.02	1.01	4.91
4.00	18.00	56.00	8.04	1.81	13.33	0.80	5.00
4.50	19.00	65.25	6.90	2.01	14.22	0.66	5.24
5.00	20.00	75.00	6.00	2.22	15.61	0.55	5.56
5.50	21.00	85.25	5.28	2.42	17.47	0.46	5.93
6.00	22.00	96.00	4.69	2.62	19.77	0.40	6.34
6.50	23.00	107.25	4.20	2.81	22.50	0.34	6.77

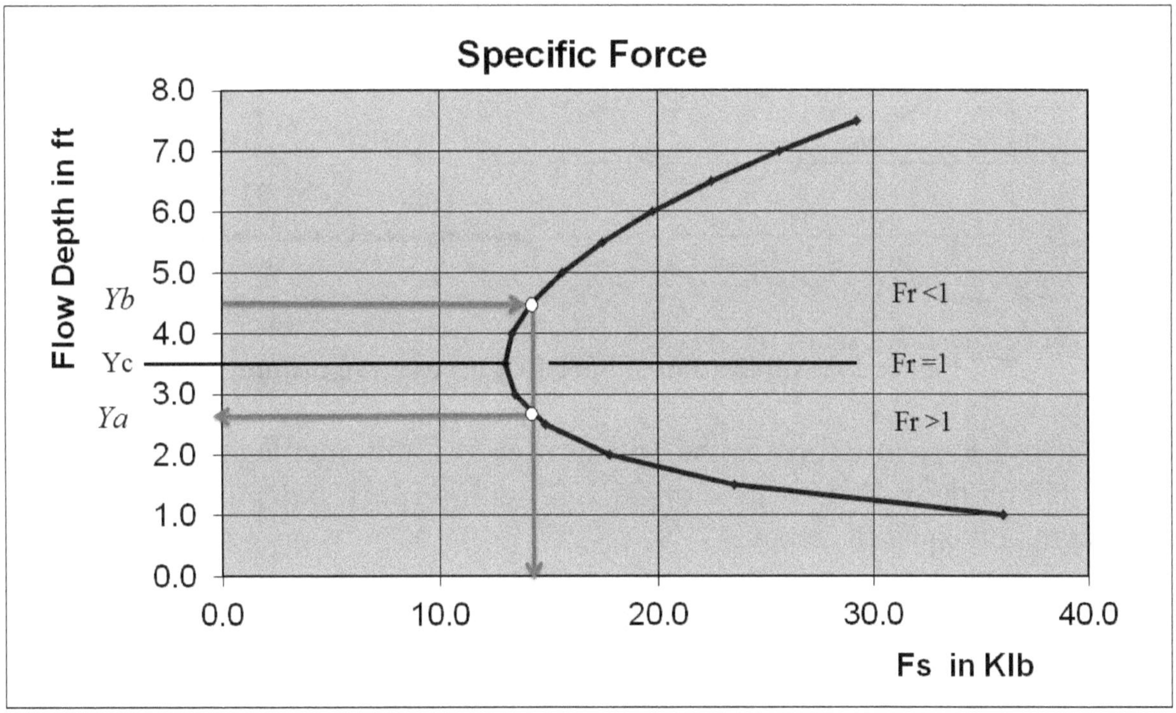

Figure 2.5. Specific Force Curve.

on a flat floor is often analyzed by the principle of momentum force. In other word, the two flow depths upstream and downstream of the jump shall have the same force as described in Eq 2.12. Such a pair of flow depths, (Y_a, Y_b), sharing the same specific force is termed *sequent depths.*

$$F_a = F_b + \Delta F \approx F_b \quad (2.19)$$

$$\Delta E = E_a - E_b \quad (2.20)$$

In which F_a= specific force at Section *a*, F_b= specific force at Section *b*, ΔF= external flow which is negligible, ΔE= energy dissipation in jump, E_a = specific energy at Section *a* and E_b = specific energy at Section b.

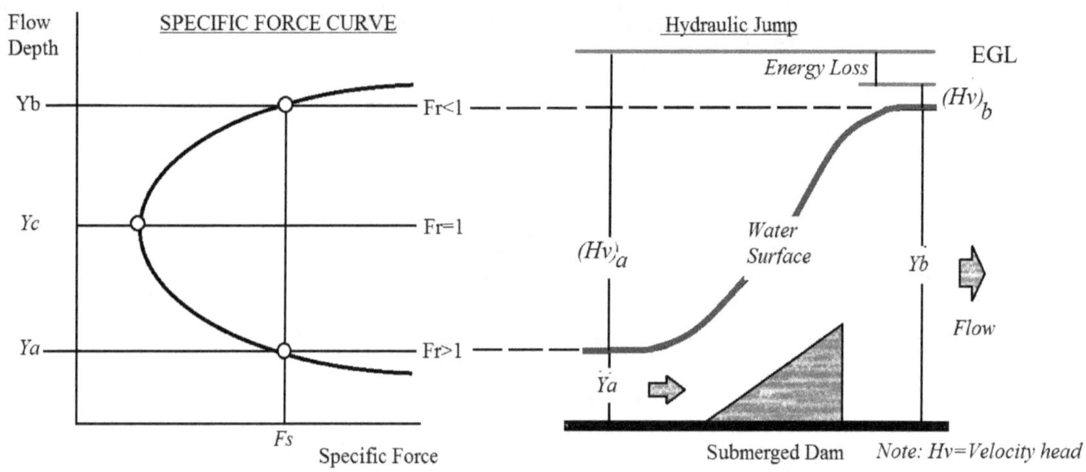

Figure 2.6. Application of Specific Force Curve to Jump.

It is noted that the specific force at the critical flow is a minimum again. This fact can be proved by taking the first derivative of Eq 2.12 with respect to depth Y, and setting the resultant equation equal to zero.

Example 2.5: The spillway in Figure 2.7 carries a flow of 450 cfs on a slope of 3.5% with Manning's N=0.015. The incoming flow upstream of the jump is set to be the normal flow, Y_a=Yn. The stilling pool backed up with a weir provides a sufficient tailwater water depth, Yt, because of the flow overtops the weir. (1) Determine the sequent depth, Y_b, downstream of the hydraulic jump (2) Estimate the energy dissipation.

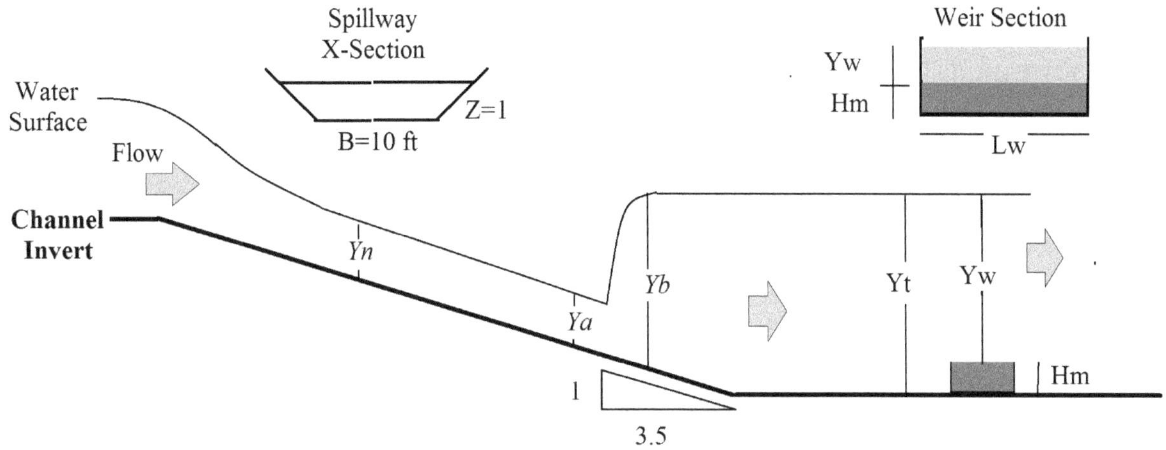

Figure 2.7. Analysis of Hydraulic Jump.

Solution:

The normal flow condition for the given condition is computed as:

Flow Depth Yn ft	Flow Area An sq ft	Wetted P-meter Pn ft	Hy-Radius Rn ft	Flow Velocity Un fps	Flow rate Q cfs	Froude Numer Fr	Sp Energy Es ft	Depth to Centroid Yh ft	Sp Force Fs klb
1.70	19.9	14.8	1.34	22.60	450.0	3.27	9.61	0.81	20.73

It is confirmed that the incoming flow is supercritical flow with Fr=3.27. It will produce a hydraulic jump when the sufficient tailwater exists. Let Y_a= *Yn*= 1.7 feet. It is found that Y_b=*6.18* feet and Fr=0.37 can produce the same specific force as 20.73 kilo-pounds.

Station Depth	Flow Depth Y ft	Flow Area A sq ft	Top Width T ft	Flow Velocity u fps	Kinetic Energy u^2/2g ft	Sp. Energy Es ft	Froude Numer Fr	Depth to Centroid Yo ft	Sp. Force Fs Klbs
Y=Ya=Yn	1.70	19.9	13.40	22.60	7.93	9.63	3.27	0.81	20.73
Y=Yb	6.18	100.1	22.37	4.50	0.31	6.50	0.37	2.69	20.73
Y=Yc	3.51	47.5	17.03	9.47	1.39	4.91	1.00	1.60	13.02

The difference in specific energy before and after the jump is calculated as:

$\Delta E_s = 9.63 - 6.50 = 3.13$ feet or pound-ft per pound of water.

$\textit{Energy Dissipation} = \Delta E_s \gamma Q = 3.13 \times 62.4 \times 450 = \textit{87.89}$ kilo-pound-ft

Hydraulic jump creates a tremendous energy loss through the plunging pool. Having known the geometry of the plunging pool, the location and length of the jump can further be determined. More details will be discussed in Chapter 3.

2.2.1. Critical Flow on Specific Force Curve

Both specific energy and force curves are centered with the critical flow for the given condition. In summary, the critical flow is characterized with:

(1) a minimum specific energy,
(2) a velocity head equal to a half of the critical hydraulic depth,
(3) a minimum specific force, and
(4) Froude number equal to unity.

As indicated in Eq 2.7, the critical flow is defined by Fr=1. For a given cross section, the flow regime varies with respect to flow depth. However, the critical flow depth is independent of channel roughness and slope.

2.2.2. Special Case for Specific Force Curve

Assuming that the external force is negligible and the value of β in Eq 2.16 is unity, Eq 2.19 is expanded for a hydraulic jump as:

$$\gamma Y_a + \rho Q A_a = \gamma Y_b + \rho Q A_b + \Delta F \quad (\Delta F=0) \tag{2.21}$$

The subscript of "*a*" are the variable associated with the "Section *a*" upstream of the jump, so is the subscript "b". For *a rectangular channel* in Figure 2.8, we have:

$$Y_o = 0.5\,Y \tag{2.22}$$

$$A = BY \tag{2.23}$$

$$U = Q/BY \tag{2.24}$$

$$F_r = \frac{U}{\sqrt{gY}} \tag{2.25}$$

Substituting the above relationships into Eq 2.2 results in:

$$\frac{Y_b}{Y_a} = \frac{1}{2}\left[\sqrt{1+8(F_r)_a^2} - 1\right] \tag{2.26}$$

Equation 2.26 depicts the relationship between sequent depths in a rectangular channel. Either the downstream or upstream depth is known, its sequent depth in a rectangular channel can be computed by Eq 2.26.

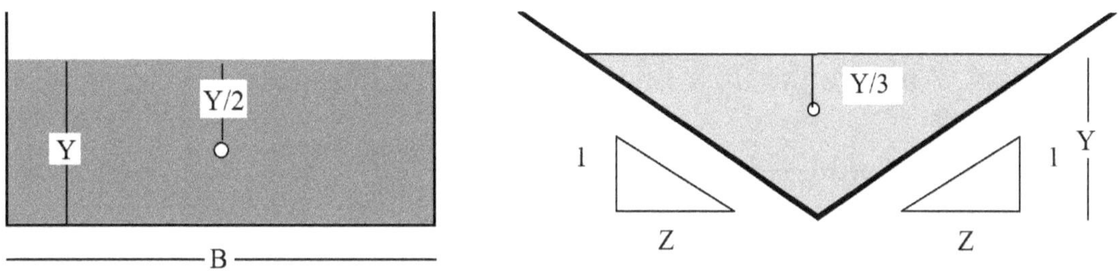

Figure 2.8. Rectangular and Triangular Channels.

For a symmetric *triangular channel* in Figure 2.8 with a side slope of *1V:ZH*, we have:

$$Y_h = Y/3 \tag{2.27}$$

$$A = ZY^2 \tag{2.28}$$

$$U = Q/(ZY^2) \tag{2.29}$$

Substituting the above equations into Eq 2.21 yields

$$\frac{1}{3}(Y_b^{\,3} - Y_a^{\,3}) = \frac{Q}{Zg}\left(\frac{1}{Y_b^{\,3}} - \frac{1}{Y_a^{\,3}}\right) \tag{2.30}$$

Equation 2.30 can be solved by trial and error if either Y_a or Y_b is given.

2.3. CLOSING

Any flow transition involves ΔE and ΔF. In this Chapter, the energy principle was simplified with ΔE=0 when solving the flows upstream and downstream of a hydraulic drop. Similarly, ΔF=0 was the assumption when solving the flows associated with a hydraulic jump. In most cases, the assumptions of either ΔE=0 or ΔF=0 is acceptable. Care must be taken when the flow conditions are not as simple as the assumption. For instance, when a hydraulic jump is triggered on a steep spillway, S_o>10%, the weight of water volume can no longer be ignored in the force analysis. The simplified method needs some adjustments for a better representation of the real flow conditions.

2.4. HOMEWORK

Q2.1: The flow in Figure HW 2.1 is carried in a rectangular channel with a bottom width of 20 feet. Analyze the flow downstream of the hump.

Case 1: Y= 10.0 feet and U= 10.0 ft/second
Case 2: Y=5.0 feet and U=10.0 ft/second
Case 3; Y=2.0 feet and U= 10.0 ft/second

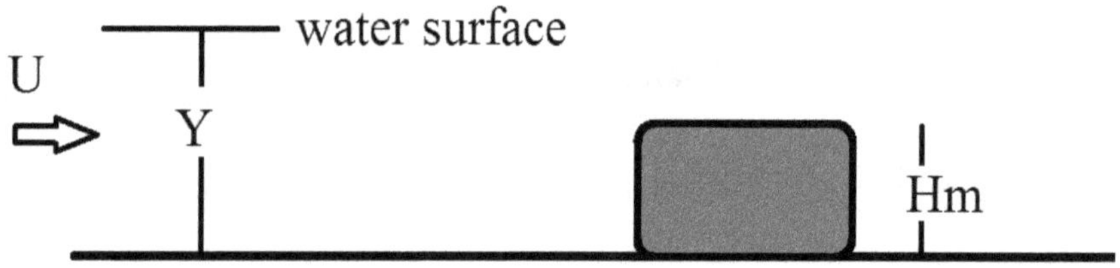

Figure HW2.1. Flow with Hump.

Q2.2: Prove that the energy loss through a hydraulic jump in a rectangular channel can be estimated as:

$$\Delta E = \frac{(Y_b - Y_a)^3}{4Y_aY_b}$$

Q2.3: Example 2.5 presents the flow analysis for the 100-yr event. Repeat the same procedure to analyze the 2-yr flow condition with Q=135 cfs on the spillway. The incoming flow upstream of the jump is set to be the normal flow, Y_a=Yn. The stilling pool backed up with a weir provides a sufficient tailwater water depth, Yt, because of the flow overtops the weir. (1) Determine the sequent depth, Y_b, downstream of the hydraulic jump (2) Locate the range of hydraulic jumps on the spillway for Q=2 to 100-yr events.

CHAPTER 3
ENERGY DISSIPATION BASIN

Hydraulic jump and drop are the most efficient measures to dissipate the excessive energy associated with a channel flow. A jump can only be triggered with an efficient tailwater depth which has to be sustained by a stilling basin while a drop needs to be encompassed within a plunging pool without scouring the downstream channel bed. In this Chapter, both theories and empirical formulas are discussed for the purpose of energy dissipation designs.

3.1. WEIR HYDRAULICS

Weir is often used across a channel for flow measurement. Flows are designed to overtop the weir and create a back water effect to produce the required tailwater condition. Weirs are classified by their cross sectional shapes, such as rectangular, triangular, and trapezoidal weirs.

3.1.1. Rectangular Weir

A rectangular weir can be as simple as a submerged dam across the channel bottom, or built with sides to contract flow. The capacity of a *rectangular weir* is determined by its crest width as:

$$Q_w = \frac{2}{3} C_d \sqrt{2g} L_e Y_w^{1.5} \qquad (3.1)$$

$$L_e = L_w - 0.1 m Y_w \qquad (3.2)$$

in which, Q_w = weir flow in [L^3/T], C_d = discharge coefficient such as 0.60 to 0.70, L_w = crest width in [L], and L_e= effective weir width in [L], g = gravitational acceleration in [L/T^2], and m = number of end contractions. Figure 3.1 presents 3 cases for m=2 for two-side contraction, m=1 for one-side contraction, and m=0 for the case with no contraction.

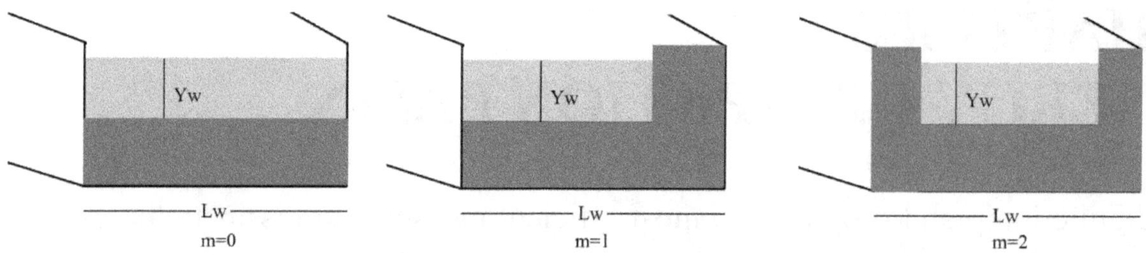

Figure 3.1. Rectangular Weir with Various End Contractions.

Equation 3.1 is dimensionally consistent. Using a lumped coefficient, Eq 3.1 is reduced to:

$$Q_w = C_w L_e Y^{\frac{3}{2}} \tag{3.3}$$

in which C_w = rectangular weir coefficient between 2.6 to 3.8 for feet-second units or 1.4 to 2.1 for meter-second units.

3.1.2. Triangular Weir

A triangular weir is built for the purpose of flow measurement. The capacity of a *triangular weir* is governed by its V-notch angle, θ, as illustrated in Figure 3.2.

$$Q_w = \frac{8}{15} C_d \sqrt{2g} \tan\left[\frac{\theta}{2}\right] Y_w^{2.5} \tag{3.4}$$

Using a lumped coefficient, Eq 3.4 is further reduced to:

$$Q_w = C_t \tan\left(\frac{\theta}{2}\right) Y^{\frac{5}{2}} \tag{3.5}$$

in which C_t = triangular weir coefficient 3.0 to 3.50 for feet-second units or 1.65 to 2.0 for meter-second units.

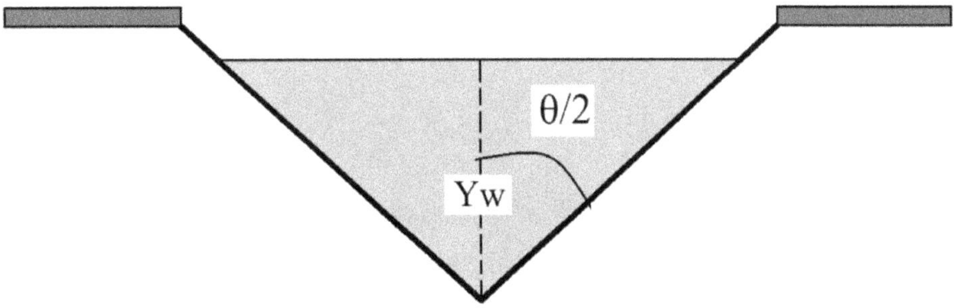

Figure 3.2. Triangular Weir.

3.1.3. Trapezoidal Weir

As illustrated in Figure 3.3, a *trapezoidal weir* is composed of a rectangular weir with a crest length equal to the trapezoidal bottom width and a triangular weir with the notch angle equal to the trapezoidal side slope, Z, as:

$$\frac{\theta}{2} = \frac{\pi}{2} - \tan^{-1}(\frac{1}{Z}) \tag{3.6}$$

In which θ = the central angle in radians for triangular weir, Z= aide slope in [L/L]. Ignoring the minor adjustment on the end contraction, the capacity of a trapezoidal weir is derived as:

$$Q_w = \frac{2}{3}\sqrt{2g}C_d\{L_w Y_w^{1.5} + \frac{4}{5}\tan\left[\frac{\theta}{2}\right]Y_w^{2.5}\} \tag{3.7}$$

Equation 3.7 does not take the internal friction into consideration. As a result, it is advisable to use a smaller value of 0.5 to 0.6 for the discharge coefficient.

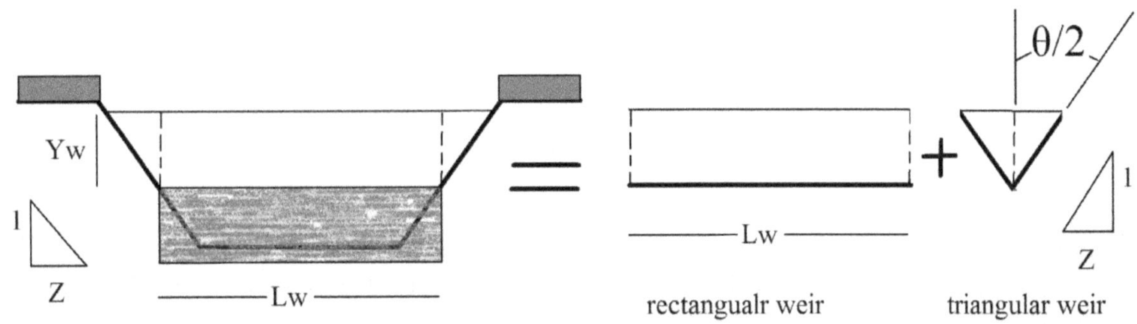

Figure 3.3. Trapezoidal Weir.

Example 3.1: A weir is installed in a trapezoidal channel as shown in Figure 3.3. Knowing L_W=18 feet, Z=1, and C_o=0.55, find the overtopping water depth, Y_w, for a discharge of 300 cfs.

Solution:

$$\frac{\theta}{2}=\frac{\pi}{2}-\tan^{-1}(\frac{1}{Z})=\frac{3.1416}{2}-\tan^{-1}(\frac{1}{1})=0.785\ radians\ (45\text{-degree})$$

$$300=\frac{2}{3}\sqrt{2\times 32.2}\times 0.55\times\{18.0\times Y_w^{1.5}+\frac{4}{5}\tan[0.785]Y_w^{2.5}\}$$

By trial and error, Y_w=2.93 *feet*

3.2. STILLING POOL

As illustrated in Figure 3.4a and 3.4b, a stilling basin is located at the downstream end of a long and steep spillway. The water flow is released from the upstream reservoir into the spillway. As expected, the supercritical flow will result in a hydraulic jump.

(a) Spillway and Stilling Basin | (b) Hydraulic Jump on Spillway

Figure 3.4. Example of Spillway and Stilling Basin.

The stilling basin as illustrated in Figure 3.5 is designed to provide adequate tailwater depth to trigger the hydraulic jump. The water pool in the basin is sustained by the backwater effect from the downstream weir. The height of the weir is chosen to meet the design criteria for the incoming design flow. The water surface profile is extended from the pool into the spillway. The water depth on the spillway changes due to the spillway's slope. An inflow will develop the normal flow on the spillway, and jump at the location where the water depth provides the required specific force associated with the incoming flow.

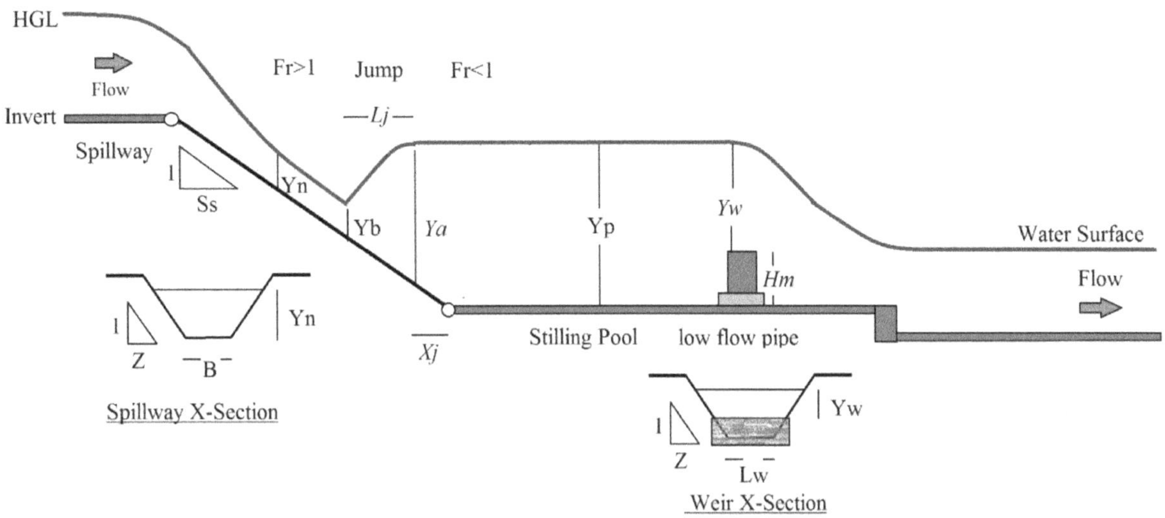

Figure 3.5. Hydraulic Jump in Stilling Pool.

3.2.1. Design Flow

Design flow is selected based on economic studies or pre-selected level of risk. The system under design shall be sized to accommodate the design flow. The design procedure shall begin with the upstream incoming flow condition.

For the given design flow, the incoming flow on the spillway is assumed to have the normal flow as:

$$Y_b = Y_n \tag{3.8}$$

Based on the momentum principle, the sequent depth, Y_a, is determined by:

$$F_a = F_b + \Delta F \quad \text{with} \quad \Delta F \approx 0 \tag{3.9}$$

Hydraulic jump involves high turbulence and energy dissipation. From the design aspect, it is preferred having the jump on the concrete spillway. The sequent depth is located at the pre-selection location as:

$$Y_P = Y_a + X_j S_s \tag{3.10}$$

Where X_j = distance from the foot of spillway in [L], and S_s= slope on spillway in [L/L]. Next, we shall investigate the weir geometry. Accept the critical depth on top of the weir. The height of the weir is:

$$Y_w = Y_C$$

$$H_m = Y_P - Y_w = Y_P - Y_C \tag{3.11}$$

Equation 3.11 is the balance of hydraulic grade line (HGL). It is an acceptable simplification of the balance of energy grade line (EGL) because the velocity head in the stilling pool is negligible. The weir is sized for the design flow. Therefore it is necessary to have one or two low-flow pipes placed on the floor to pass the frequent flows. The length of the pool mainly depends on the length of hydraulic jump which is highly empirical. Based on USBR studies, the jump length is approximated as:

$$\frac{L_j}{Y_b} = 10(Fr_b - 1) \tag{3.12}$$

Where L_j= length of jump in [L]. Equation 3.12 only represents the jump length. In practice, a safety factor of 1.25 to 1.50 shall be considered.

Example 3.2: Referring to Figure 3.5, the concrete spillway has B=10 feet, Z=1, and S_s=5%. The design flow is 600 cfs. Design the stilling pool to have the jump on the spillway.

Solution:

1. Determine the Incoming flow condition on the spillway.

Flow Depth Yn ft	Flow Area An sq ft	Wetted P-meter Pn ft	Hy-Radius Rn ft	Flow Velocity Un fps	Flow rate Q cfs	Froude Numer Fr	Sp Energy Es ft	Depth to Centroid Yh ft	Sp Force Fs klb
1.82	21.4	15.1	1.42	27.98	600.0	3.93	13.93	0.86	33.72

2. Set Y_b= Y_n= 1.82 feet, Fr= 3.93 and Fs= 33.72 kilo-pounds.

3. Find the Sequent Depth, Y_a=7.79 feet, based on *Fs*=33.72 kilo-pounds

Station Depth	Flow Depth Y ft	Flow Area A sq ft	Top Width T ft	Flow Velocity u fps	Kinetic Energy u^2/2g ft	Sp. Energy Es ft	Froude Numer Fr	Depth to Centroid Yo ft	Sp. Force Fs Klbs
Y=Yb=Yn	1.82	21.4	13.63	27.98	12.16	13.97	3.93	0.86	33.72
Y=Ya	7.79	138.6	25.58	4.33	0.29	8.08	0.33	3.32	33.72
Y=Yc	4.16	59.0	18.32	10.18	1.61	5.77	1.00	1.87	18.74

4. Design the Weir Section Based on Critical Flow Condition.

Set Xj=10 feet upstream from the toe of the spillway.

$$Y_P = Y_a + X_j S_s = 7.79 + 10 \times 0.05 = 8.29 \text{ feet}$$

$$Y_w = Y_C = 4.16 \ ft$$

$$H_m = Y_P - Y_C = 8.29 - 4.16 = 4.13 \ feet$$

For this case, one 6-inch circular pipe shall be installed as the low flow pipe to pass frequent flows.

5. Estimate the length of the jump.

$$\frac{L_j}{Y_b} = 10(Fr_b - 1) = 10(3.93 - 1) = 29.3 \text{ or Lj=53.3 feet}$$

3.2.2. Operational Flow

After the system is built, a wide spectrum of water flows will go through the system. With the as-built structures, the flow in operation will accommodate the system. The analysis of the backwater surface profile shall start from the downstream control section which is the weir section.

Aided with Eq's 3.1 and 3.3, the overtopping depth at the weir section can be determined. As a result, the water depth in the stilling pool is:

$$Y_p = Y_w + H_m \tag{3.13}$$

Next, the balance of specific forces for the flow in operation provides the pair of sequent depths, (Y_a, Y_b). The location of the jump on the spillway is determined as:

$$X_j = \frac{Y_o - Y_b}{S_S} \tag{3.14}$$

Example 3.3: Continue with Example 3.2. Referring to Figure 3.5, the concrete spillway has B=10 feet, Z=1, and S_s=5%. Analyze the hydraulic jump for a flow of 300 cfs.

Solution:

1. Determine the normal flow on the spillway.

Flow Depth Yn ft	Flow Area An sq ft	Wetted P-meter Pn ft	Hy-Radius Rn ft	Flow Velocity Un fps	Flow rate Q cfs	Froude Numer Fr	Sp Energy Es ft	Depth to Centroid Yh ft	Sp Force Fs klb
1.20	13.5	13.4	1.01	22.24	300.0	3.76	8.86	0.58	13.43

2: Set Y_b=Y_n. Find Ya based on *Fs*=13.43 kilo-pounds.

Station Depth	Flow Depth Y ft	Flow Area A sq ft	Top Width T ft	Flow Velocity u fps	Kinetic Energy u^2/2g ft	Sp. Energy Es ft	Froude Numer Fr	Depth to Centroid Yo ft	Sp. Force Fs Klbs
Y=Yb	1.92	22.9	13.84	13.11	2.67	4.59	1.80	0.91	8.92
Y=Ya	3.78	52.1	17.56	5.76	0.52	4.29	0.59	1.71	8.92

3. Analyze the flow of 300 cfs on the as-built weir.
According to Example 3.1, Y_w=2.93 feet

4. Determine the water depth in the pool.
Knowing that the height of the as-built weir is 4.13 feet from Example 3.2, the water depth in the pool is:

$$Y_p = Y_w + H_m = 2.93 + 4.13 = 7.06 \text{ feet}$$

5. Locate the hydraulic jump on the spillway.

$$X_j = \frac{Y_o - Y_b}{S_S} = \frac{7.06 - 3.78}{0.05} = 65.5 \text{ feet upstream of the toe.}$$

6. Determine the length of the jump.

$$\frac{L_j}{Y_b} = 10(Fr_b - 1) = 10(1.8 - 1) = 8.0 \text{ or } Lj = 15.4 \text{ feet}$$

It is important to understand that the stilling system is designed for the design flow, but it has to be tested for a range of frequent flows to make sure the system can pass all flows safely.

3.3. PLUNGING POOL

A plunging basin is located at the downstream end of a drop structure. Drop is an efficient measure to dissipate the flow energy.

(a) Plunging Pool in Channel | (b) Jump in Plunging Pool

Figure 3.6. Examples of Plunging Pool.

A drop is characterized with its drop height, h, as shown in Figure 3.7. The unit-width flow overtops the drop structure and plunges into the pool. The water jet produces a parabolic trajectory to impinge on the pool floor.

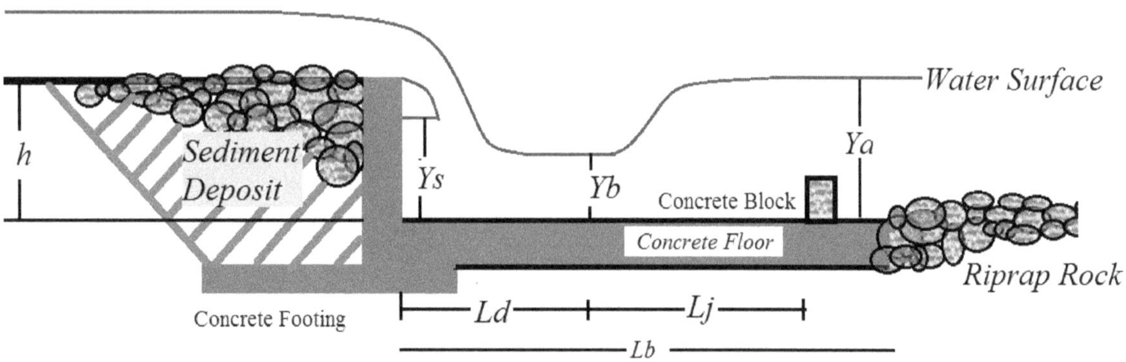

Figure 3.7. Plunging Pool.

The geometry of this jet flow is described by the drop height and drop number that are defined as:

$$q = \frac{Q}{T} \tag{3.15}$$

$$D_n = \frac{q^2}{gh^3} \tag{3.16}$$

in which Q= design flow in [L^3/T], q= unit-width discharge in [L^2/T], T= top width in [L] of weir crest, D_n=drop number in [L^0], g= gravitational acceleration in [L/T^2], and h= drop height in [L]. The drop-pool section is designed to have a rectangular cross section with its width equal to the top width, T, of the drop structure.

$$\frac{L_d}{h} = 4.30 D_n^{0.27} \tag{3.17}$$

$$\frac{Y_S}{h} = 1.00 D_n^{0.22} \tag{3.18}$$

$$\frac{Y_b}{h} = 0.54 D_n^{0.425} \tag{3.19}$$

$$\frac{Y_a}{h} = 1.66 D_n^{0.27} \tag{3.20}$$

$$\frac{L_j}{Y_b} = 10\left(\frac{U_b}{\sqrt{gY_b}} - 1\right) \tag{3.21}$$

$$L_b \geq (L_j + L_d) \tag{3.22}$$

In which L_d = distance in [L] from drop wall to depth Y_b, Y_S= pool depth in [L] near drop wall, Y_b = pool depth in [L], U_b= average flow at Section b in [L/T], Y_a = tailwater depth in [L], L_j= jump length in [L], and L_b= length of pool in [L].

Example 3.4: *Given* Q=1000cfs, T=60.83 ft, and h= 3.79 feet. Design the dimension of the plunging pool.

Drop height per reach	*h*=	3.79	*ft*
Unit flow rate per foot	*q*=*Q*/*T*=	16.44	*cfs/ft*
Drop number	*Dn*=	0.15	
Water depth in pool	Y_S=	2.51	*ft*
Water depth before jump	Y_b=	0.92	*ft*
Water depth after jump	Y_a=	3.80	*ft*
Location of jet impingement	L_d=	9.84	*ft*
Length of jump	L_j =	20.88	*ft*
Minimal total length of pool	L_d+L_j=	30.71	*ft*

3.3.1. Low-Flow Weir on Top of Drop Structure

From the point view of safety, a drop structure in Figure 3.8a is always sized to withstand the extreme event. However, its daily operation is to pass frequent, small flows. As a result, the top weir should be designed with a low flow notch. The low flow rate is approximately 1 to 3% of the design flow such as the 100-yr peak flow. Applying the weir flow formula to the low flow notch in Figure 3.8b for the low flow rate yields:

$$Q_L = \frac{2}{3} C_d \sqrt{2g} W_1 H_1^{\frac{3}{2}} \tag{3.23}$$

where Q_L =low flow in [L^3/T], W_1= notch width in [L], and H_1= low head depth in [L] on notch weir. In practice, the low head depth is pre-selected from the range of 1 to 2 feet. The only unknown, W_1, in Eq 3.23 can be determined. Under the design flow, the high head depth on top of the drop structure applies to both the notch weir in the center, and two overflow weirs on both sides. The weir flow formula for the composite weir section under the high head depth is written as:

$$Q_H = \frac{2}{3} C_d \sqrt{2g} [W_1 (H_1 + H_2)^{\frac{3}{2}} + (W_2 - W_1) H_2^{\frac{3}{2}}] \tag{3.24}$$

where Q_H= design flow [L^3/T] under high head depth, W_2 = top width in [L] of the drop structure, and H_2= water depth in [L] on top of overflow weir. In practice, W_2 in [L] is set to be the top width of the approaching channel section upstream of the drop structure. As a result, the only unknown, H_2, in Eq 3.24 can be solved for the design flow.

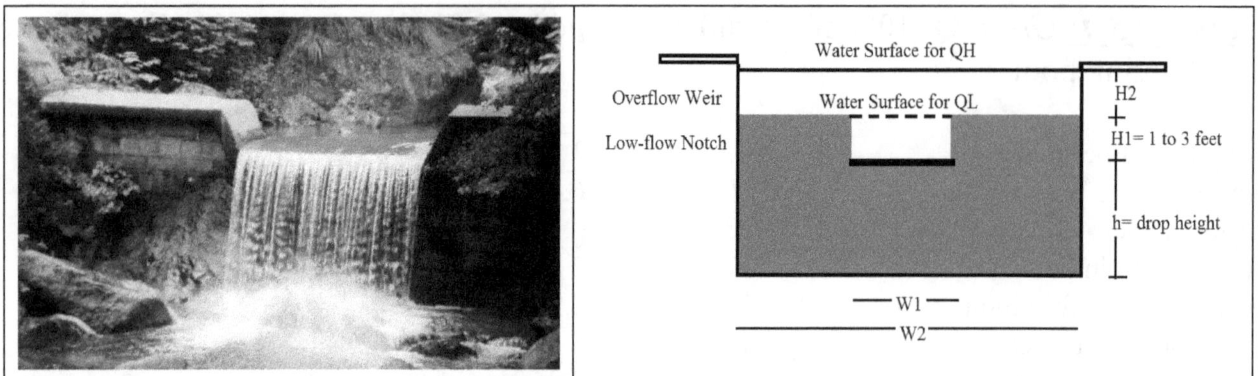

(a) Front View | (b) Low Flow Notch

Figure 3.8. Weir Section On Top of Drop Structure.

Example 3.5: The trapezoidal channel in Figure 3.8(b) is designed with Q_L=20 cfs, Q_H=1000 cfs, H_1= 1.0 ft and C_o=0.6 or C_W=3.21 using feet-second unit. The top width of the channel flow is 60.83 ft. Size the dimensions of the weir section on top of the drop structure.

Solution:

$$Q_L = \frac{2}{3} C_d \sqrt{2g} W_1 H_1^{\frac{3}{2}} = \frac{2}{3} \times 0.6 \times \sqrt{2 \times 32.2}\, W_1 (1)^{3/2} = 20$$

So, *W_1=6.23 ft. Let W2=60.83 ft*

$$Q_H = \frac{2}{3} \times 0.60 \sqrt{2 \times 32.2} [6.23 \times (1 + H_2)^{\frac{3}{2}} + (60.83 - 6.23) H_2^{\frac{3}{2}}]$$

So, H_2=2.86 ft

3.4. HOMEWORK

Q3.1: A riprap trapezoidal channel is to be built on a slope of 0.5% with a bottom width of B=20 feet, side slope of Z=1 or 1V:1H, and roughness of N=0.045. The design discharge is Q=2000 cfs. A shallow check dam is proposed to be installed across the channel floor. Determine the height of the check dam.

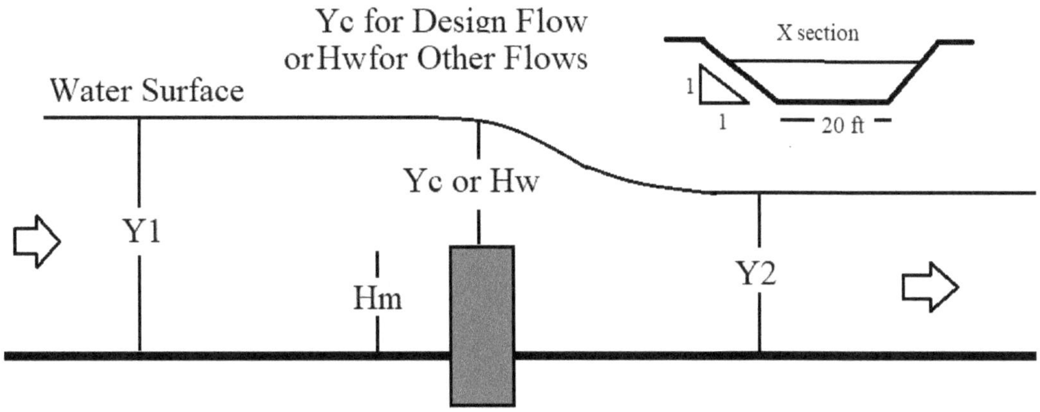

Figure HW3.1. Design of Check Dam.

Solution: The normal flow condition is tabulated as:

Flow Depth Y Ft	Flow Area A sq ft	Wetted P-meter P ft	Hy-Radius R ft	Flow Velocity U fps	Froude Number Fr	Specific Energy Es ft
9.09	264.3	45.7	5.78	7.57	0.51	9.97

This is a new structure to be built. It is necessary to avoid any backwater effect. As a result, the critical flow, Yc, is designed to be on top of the proposed dam. As a result, the height of the check dam shall be 1.43 feet.

Station	Flow Depth Y ft	Flow Area A sq ft	Top Width T Ft	Flow Velocity U Fps	Kinetic Energy $u^2/2g$ ft	Sp. Energy Es ft	Froude Number Fr
Y=Y1	9.09	264.3	38.17	7.57	0.89	9.98	0.51
Y=Y2	4.30	104.6	28.61	19.12	5.68	9.98	1.76
Y=Yc	6.08	158.7	32.17	12.60	2.47	8.55	1.00
The height of Check Dam=			1.43	Feet	Es2=Esc + Hm		

Q3.2: Having the check dam built across the channel in Q3.1. Determine the weir flow across the check dam when the flow is 500 cfs.

Solution: Assuming the same cross section is applied to the weir top, the weir flow is analyzed as:

$B = 20\ ft$ and $\theta = 45^0$

$$Q_w = \frac{2}{3}\sqrt{2g}C_d\{BH_w^{1.5} + \frac{4}{5}\tan\left[\frac{\theta}{2}\right]H_w^{2.5}\} = 500\ cfs$$

By trial and error, find the overtopping depth, H_w

Q3.3: A concrete trapezoidal chute is to be built on a slope of 5.0% with a bottom width of 20 feet, side slope of 1V:1H, and roughness of N=0.014. The design discharge is 500 cfs. A stilling basin is to be formed with a crossing weir. (a) Determine the location of jump, tailwater depth, and height of the weir for the design flow of 500 cfs. (b) Having the system built, analyze the flow of 50 cfs through the system.

Solution: All depths are in feet.

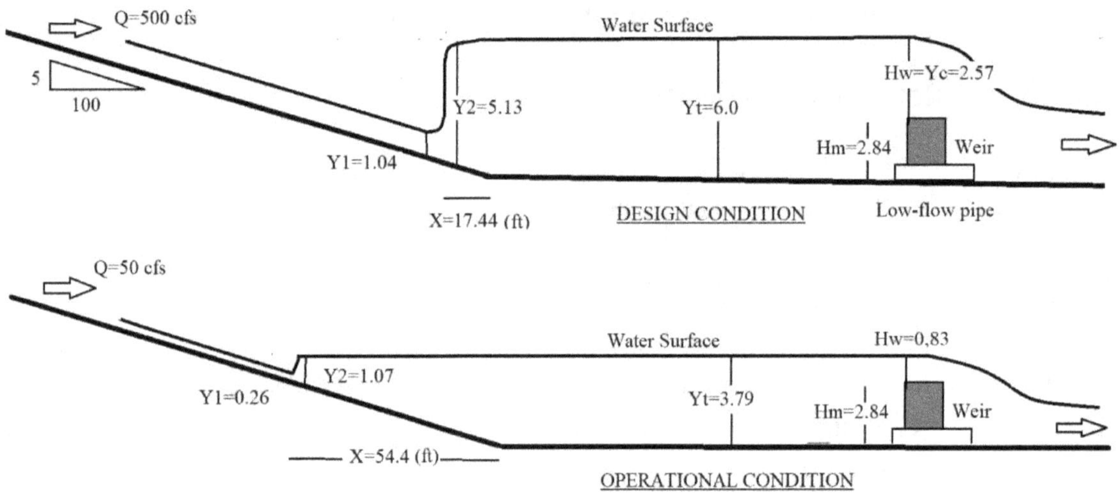

Figure HW 3.2. Design of Stilling Basin.

CHAPTER 4
DROP STRUCTURE FOR GRADE CONTROL

Drop structure is often employed to control the energy grade line (EGL) along a waterway. A drop creates a discontinuity on the channel invert and efficiently dissipates the excessive flow energy through impingements on the floor. When the flow line (thalweg) of a channel is too steep for the design condition, erosion and scour may occur to the channel bed and embankments. To mitigate sediment aggradation and degradation problems, a drop structure is installed across a channel to create a vertical drop or a short sloping drop. In this Chapter, the concepts of permissible velocity and Froude number are introduced as design criteria to prevent the potentials of streambed erosion and embankment scours.

(a) Vertical Drop Structure | (b) Sloping Drop Structure

Figure 4.1. Examples of Drop Structures.

4.1. CONCEPT OF PERMISSIBLE VELOCITY

Sediment transport in a natural or grass-lined waterway is closely related to the flow velocity. The higher the flow velocity, the more the sediment loads. The concept of permissible velocity is to accept the flow velocity up to 5 ft/sec for sandy to loamy stream bed (Type A and B Soils), or 7 ft/sec for clay stream bed (Type C and D Soils). According to Manning's formula, the permissible velocity is achieved by a grade control implemented on the channel invert using drops. With a drop, the proposed new slope for the channel invert is determined using the procedure in Figure 4.2a.

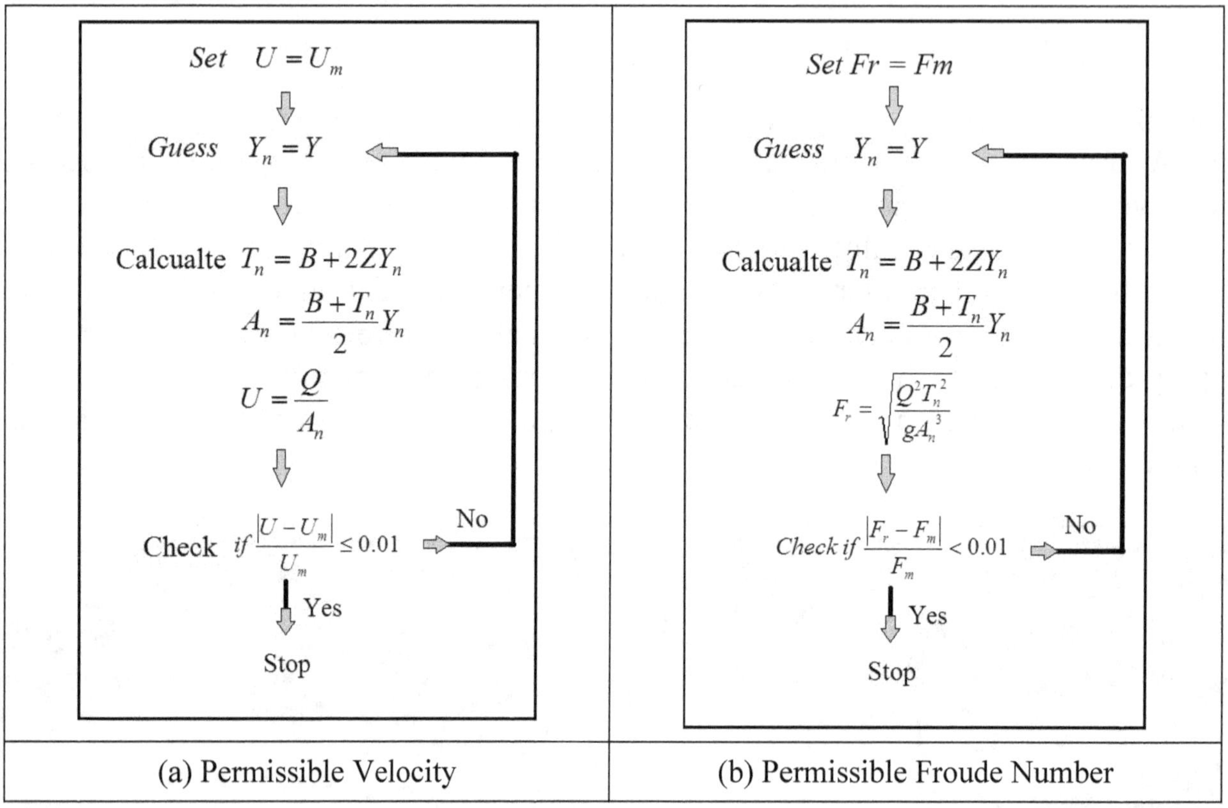

Figure 4.2. Procedure for New Slope for Permissible Velocity.

As illustrated in Figure 4.2, set the target for the flow velocity to find the corresponding flow depth. The proposed channel slope is further determined as:

$$S_{new} = \frac{N^2 U_m^2}{k_N^2 R^{4/3}} \tag{4.1}$$

$$h = (S_o - S_{new})L \tag{4.2}$$

Where S_{new} = proposed new slope in [L/L], k_N=1.486 for feet-second units or 1.0 for meter-second units, N= manning's roughness, R= hydraulic radius in [L], h= drop height in [L], S_o= existing slope in [L/L], and L = length of reach in [L]. Waterways in urban areas area are often accessible for the public because they serve as water parks and open space. For safety, the drop height is recommended not to exceed 5 feet. A given channel as shown in Figure 4.3, a series of drops is built and each drop controls a reach of 100 to 300 feet.

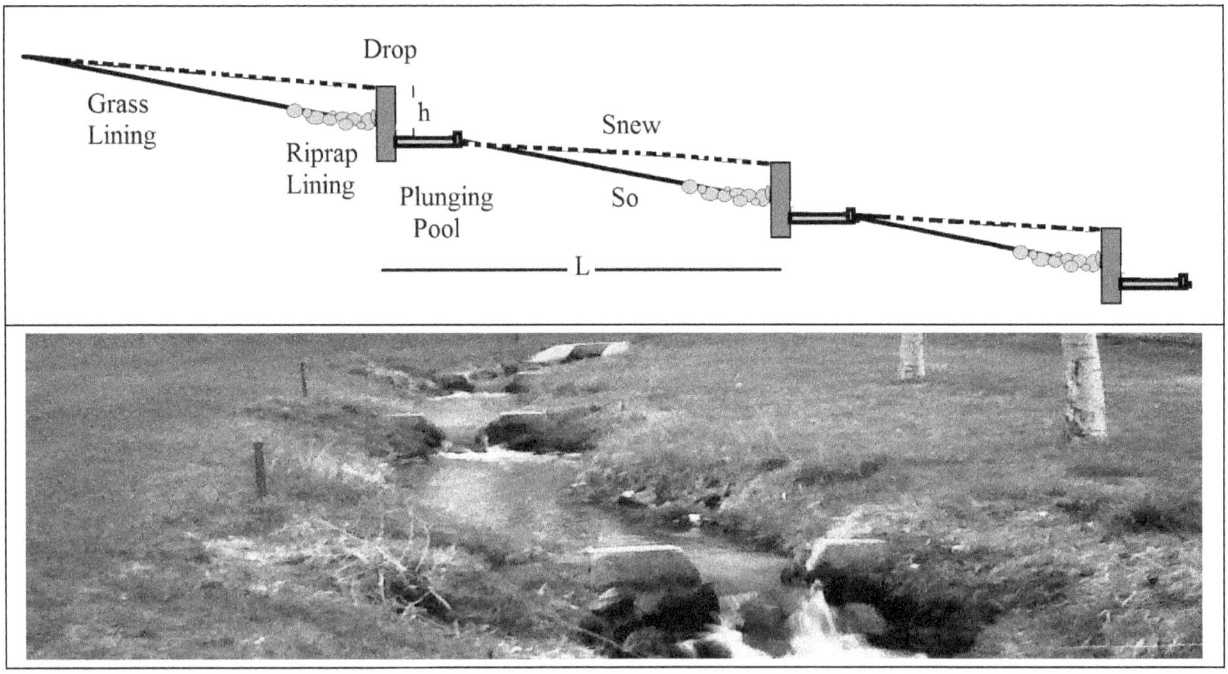

Figure 4.3. Continuous Drops.

Example 4.1: A channel is to be constructed on Type C soils. The channel will carry a design flow of 2000 cfs in a grass-lined trapezoidal cross section with bottom width B= 15 feet on a slope of S_o=3%. Determine the drop height and reach length.

Solutions: A Type C soil is clay in nature. Set U_m=7.0 ft/sec and Z=4 and N= 0.035 for grass lining.

Guess $Y=Y_n= 6.78\,ft$

$$T = B + 2ZY = 15.0 + 2\times 4\times 6.78 = 69.26 \quad \text{feet}$$

$$A = (B+T)Y/2 = 285.7 \quad \text{sq ft}$$

Check $U = Q/A = 7.0$ ft/sec (Ok! Y_n=6.78 ft is accepted.)

$$F_r = \sqrt{\frac{Q^2T^2}{gA^3}} = \sqrt{\frac{2000^2\times 69.26^2}{32.2\times 285.7^3}} = 0.61(<1) \quad \text{subcritical flow}$$

$$P = B + 2\sqrt{1+Z^2}Y = 70.93 \text{ feet}$$

$$R = A / P = 4.03 \text{ feet}$$

$$S_{new} = \frac{N^2 U_m^2}{k_N^2 R^{4/3}} = \frac{0.035^2 \times 7.0^2}{1.486^2 \times 4.03^{4/3}} = 0.0042 \text{ ft/ft}$$

$$h = (S_o - S_{new})L = (0.03 - 0.0042) \times 180 = 4.64 (\leq 5) \textit{ feet} \text{ for a reach of 180 feet}$$

For this case, let L=180 feet and h=4.64 feet.

4.2. CONCEPT OF PERMISSIBLE FROUDE NUMBER

A drop often involves a change in flow regime in which a supercritical flow is converted into a subcritical flow. Fluctuations on water surface in a critical flow are sensitive to any changes in flow depth. An unstable flow tends to have its Froude Number in the range of 0.8≤Fr≤1.2. Therefore, it is advisable to set the permissible Froude Number to be 0.7 to 0.8. Using the Froude Number as a basis, the drop height is quantified by the permissible Froude Number. The proposed new slope for the channel invert is determined using the procedure in Figure 4.2b.

Example 4.2: Repeat Example 4.1 with a permissible Froude Number of 0.80. Determine the drop height and reach length.

Solutions: Set Fr_m=8.0. Accept Z=4 and N= 0.035 for grass lining.

Guess $Y=Y_n= 5.93\ ft$

$$T = B + 2ZY = 15.0 + 2 \times 4 \times 5.93 = 62.46 \text{ feet}$$

$$A = (B+T)Y/2 = 229.75 \text{ sq ft}$$

$$U = Q / A = 8.71 \text{ ft/sec}$$

Check $F_r = \sqrt{\frac{Q^2 T^2}{g A^3}} = \sqrt{\frac{2000^2 \times 62.46^2}{32.2 \times 229.75^3}} = 0.80 (< 1)$ subcritical flow

$P = B + 2\sqrt{1+Z^2}Y = 63.92$ feet

$R = A/P = 3.59$ feet

$$S_{new} = \frac{N^2 U_m^2}{k_N^2 R^{4/3}} = \frac{0.035^2 \times 8.71^2}{1.486^2 \times 3.59^{4/3}} = 0.0076 \text{ ft/ft}$$

$h = (S_o - S_{new})L = (0.03 - 0.0076) \times 200 = 4.48$ feet per 200 feet (close to 5 feet)

For this case, let L=200 feet and h=4.48 feet.

In comparison, Example 4.1 results in U=7.0 ft/sec and Fr=0.61 (<0.8, Ok) and Example 4.2 suggests that Fr=0.8 and U=8.71 ft/sec (>7.0 ft/sec, No!). Therefore, it is concluded that this case is dictated by the permissible velocity, and S_{new}=0.0042 and h=3.87 ft for every 150 feet of channel are recommended.

4.3. DESIGN CONSIDERATION

Drop structures can be built with gabions, sheet piling, concrete walls with footings, and riprap. For urban drainage design, it is more desirable to use a series of low head drops than a high head drop. Drop structures are susceptible to bank and bottom erosion failures, including piping from seepage, flow bypassing due to excess storm runoff, degradation and scour on channel bed and embankments. In many cases, additional bank and bottom protection will be needed for a drop structure because it induces abrupt changes to the flow. The foundation material must provide required supporting strength to resist sliding force and overturning moment. Design considerations for a drop structure include:

(1) The Length of Drop

The height of a vertical drop is limited to five feet because of expensive construction costs on retaining walls and footings. A vertical drop is directly connected to the plunging pool. Safety is always a major concern in the design of drop structure. A vertical drop in Figure 4.4 is economic, but it creates a big roller at the plunging pool. A sloped drop in Figure 4.4 is built as a chute with concrete baffles that create continuous tumbling flows into the plunging pool. The length of a sloping drop depends on the slope ranging from 10% to 20%. In comparison, sloped drop is preferred in river-walk parks.

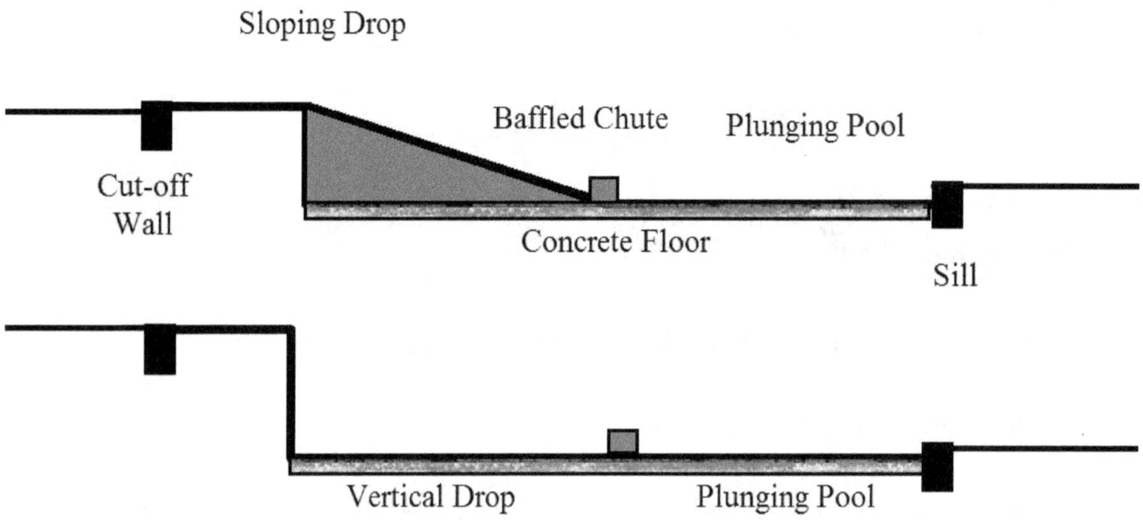

Figure 4.4. Sloping Drop and Vertical Drop.

Examples of baffled chute, grouted sloping drop, vertical drop and soil cement drop are presented in Figure 4.5.

Figure 4.5. Various Types of Drops.

(2) The Width and Height

The width of a drop structure has to be wide enough to pass the design discharge. For a riprap drop, the unit discharge shall not exceed 35 cfs per foot. The height of a drop structure is governed by the degree of protection, available construction material, required structural stability, and cost. Riprap drops built across channel floors are most economical, but not stable for a drop more than four feet because of its structural instability. Vertical drop is limited to five feet because of expensive construction costs on retaining walls and footings.

(3) The Selection of Building Materials

Examinations on flow hydraulics, channel morphologic, and soil gradations and sizes on channel bed provide some guideline on the selection of building materials. In general, concrete and grouted riprap linings as shown in Figure 4.6 are highly recommended. Soil cement can be a good substitute for an economic reason. Gabions are not recommended for a permanent drop structure because of fast deterioration.

Figure 4.6. Various Building Materials for Drops.

(4) The Forces on the Structure
Hydraulic forces, F_1 and F_2 in Future 4.7, are linearly distributed with respect to water depth. The difference between the horizontal forces may cause a sliding failure on the drop structure. The turning moment caused by the weight of the structure, horizontal forces, uplift forces, and friction force may cause a rotating failure. It is important to design the geometry of the drop structure to balance the forces and moment to avoid the failures under the critical condition. Considerations include the range of flow rates, water depths and velocities, head difference, storage upstream of the submerged dam, seepage and uplift forces, and channel flow regime, etc. Grouting and concrete are always good to improve the strength of a riprap drop structure, but an impervious surface requires an under-drain system in order to relieve uplift force due to seepage flows.

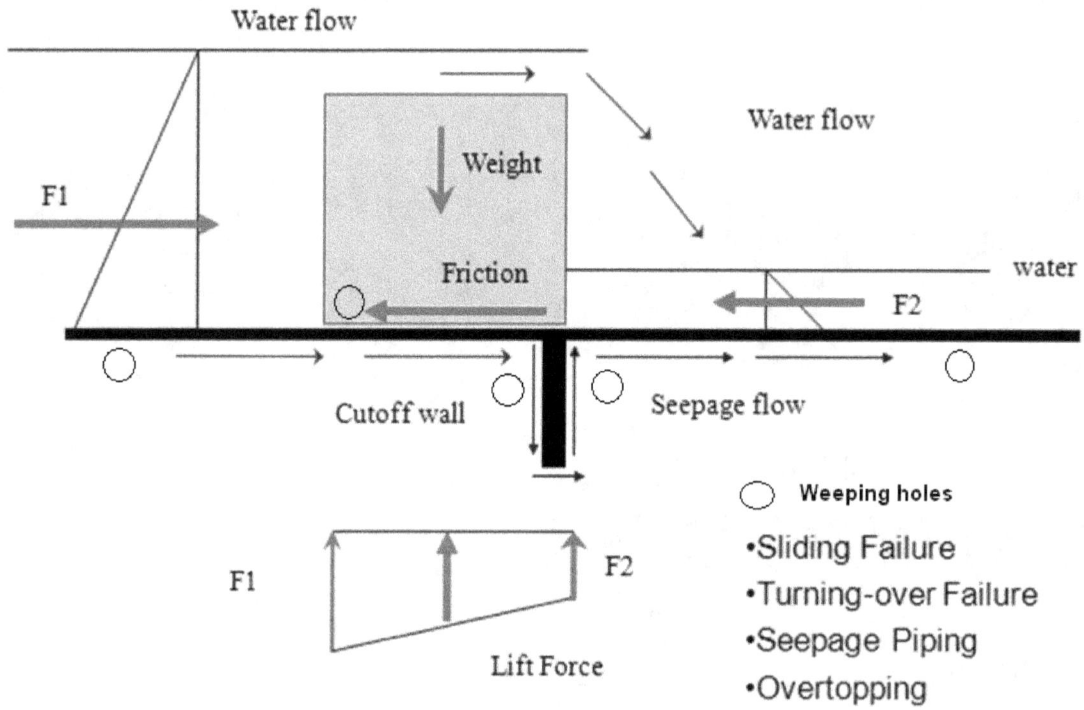

Figure 4.7. Forces Around Drop Structure.

4.4. PLUNGING POOL

A vertical drop structure produces a turbulent jet with a parabolic trajectory to impinge on the downstream channel floor. The geometry of this jet flow shown in Figure 4.8 can be described by the drop height, h, unit-width discharge, q, and drop number, D, defined as:

$$q = \frac{Q}{T} \tag{4.3}$$

$$D_n = \frac{q^2}{gh^3} \tag{4.4}$$

in which Q= design flow in [L^3/T], q= unit-width discharge in [L^2/T], T= top width in [L] of weir crest, D_n=drop number in [L^0], g= gravitational acceleration in [L/T^2], and h= drop height in [L]. The geometry of plunging pool is empirically related to the drop number as described in Section 3.3 in Chapter 3 and illustrated in Figure 4.8 in which L_d = distance in [L] from wall to depth Y_b in Figure 4.1, Y_S = pool depth in [L] near wall, Y_1 = pool depth in [L] at the beginning of the jump, Y_a =tailwater depth in [L].

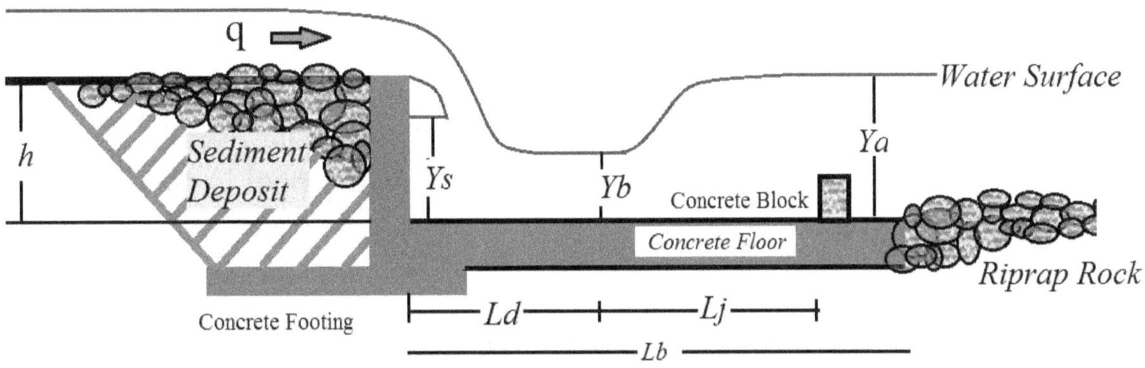

Figure 4.8. Flow over Drop Structure.

Example 4.3: A trapezoidal channel is designed to carry a discharge of 500 cfs with B= 5 feet, Z=4, and S_o= 0.02 and N=0.035. To protect the channel from erosion, the normal flow velocity for the design discharge has to be reduced to 5.0 ft/sec or below, and the flow Froude number shall be reduced to 0.60 or below. To satisfy the above criteria, a drop structure is to be installed as a grade control for each reach of 200 feet.

Solution: To design this drop structure, we have to first understand the existing flow condition. Substituting the cross sectional parameters into Manning's formula to produce the existing flow condition as:

$Y_n = 3.17\ ft,\ U_n = 8.91\ fps,$ and $Fr=1.16$

Obviously, both the Froude number and flow velocity are higher than the permissible. In order to satisfy the criteria, the proposed flow velocity is set to be:

$U_{new} = 5.0\ fps$

For Q=500 cfs, the proposed flow area becomes:

$$A_{new} = \frac{Q}{U_{new}} = 100 = Y(B+ZY) = Y(5+4Y)$$

The formula for the proposed flow area yields a solution: $Y= 4.41\ feet$. Next, substituting the proposed velocity and flow depth into Manning's formula, the only unknown is the slope for the proposed channel. The new slope for this case is found to be 0.0043. As illustrated in Figure 4.9, to reduce the channel slope from 0.02 to 0.0043 for a reach of 200 feet, the drop height is:

$h = (0.02\text{-}0.0043)*200 = 3.14\ feet$ per 200 feet

Immediately upstream of the drop structure needs a riprap protection on the existing ground. The proposed new slope is formed with the fill by water-sediment deposit before the peak flow comes during a storm event. The plunging pool is placed immediately downstream of the drop structure.

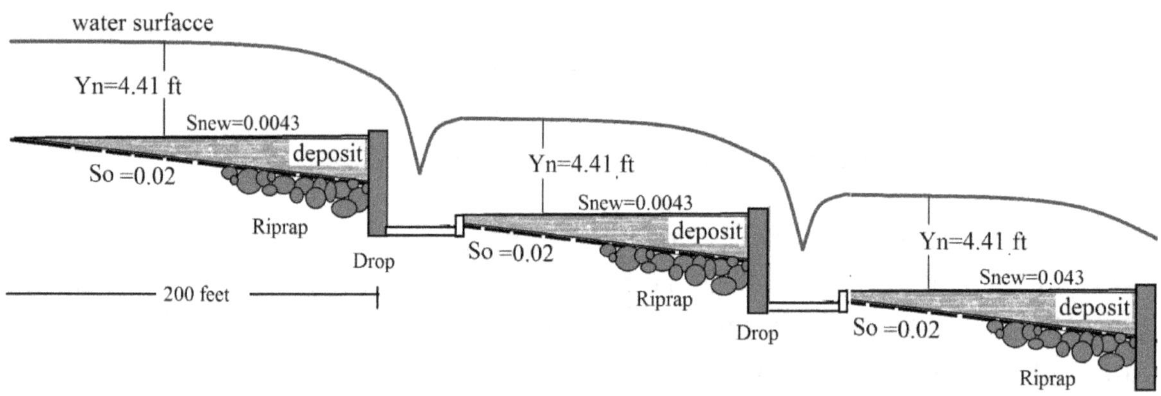

Figure 4.9. Layout of Drop Structure.

Table 4.1. Design of Drop Structure.

Design Criteria			
Max Flow Velocity	U-max=	5.00	fps
Limiting Froude Number	Fr-limit=	0.60	
Given Design Information:			
Bottom Width	B=	5.00	feet
Left Side Slope	Z1=	4.00	ft/ft
Right Side Slope	Z2=	4.00	ft/ft
Manning's n	N=	0.035	
Longitudinal Slope	S=	0.0200	ft/ft
Design Flow	Q=	500.00	cfs
Existing Normal Flow Condition			
Normal Flow Depth	Yn=	3.17	feet
Wetted Perimeter	P=	31.16	ft/ft
Top Width	T=	30.37	feet
Normal Flow Area	An=	56.10	sq ft
Hydraulic Radius	R=	1.80	feet
Flow Velocity	U=	8.91	fps
Froude Number	Fr=	1.16	
Check Difference In Q	dQ=	0.00	ft/ft
Proposed Channel Condition			
Proposed Manning's N	N-new=	0.0350	
Normal Flow Depth	Y-new=	4.41	feet
Wetted Perimeter	P-new=	41.40	ft/ft
Top Width	T-new=	40.31	ft
Normal Flow Area	A-new=	99.99	sq ft
Hydraulic Radius	R-new=	2.42	feet
Check: Flow Velocity	U-new=	5.00	fps
Check: Froude Number	Fr-new=	0.56	
New Slope	S-new=	0.0043	ft/ft
Geometry of Drop			
Reach Length	L =	200	feet
Drop Height per Reach	h=	3.14	ft
Unit Flow rate per foot	q=	22.07	cfs/ft
Drop number	D=	3.88	
Pool Length	Ld=	9.76	ft
Depth in Pool	Ys=	2.12	ft
Depth before Jump	Yb=	1.51	ft
Depth after Jump	*Ya*=	3.77	ft

4.5. WATER SURFACE PROFILE

A backwater or drawdown water surface profile is created by the downstream drop. For instance, on the new slope, the design flow is converted into a subcritical flow that will go through its critical depth on top of the drop. Between the critical depth and the normal depth upstream is termed the M-2 curve. Similarly, for a flow other than the design flow, the drop structure acts like a weir. The headwater developed on the top of the drop will back up the flow to form an M-1 curve upstream of the drop.

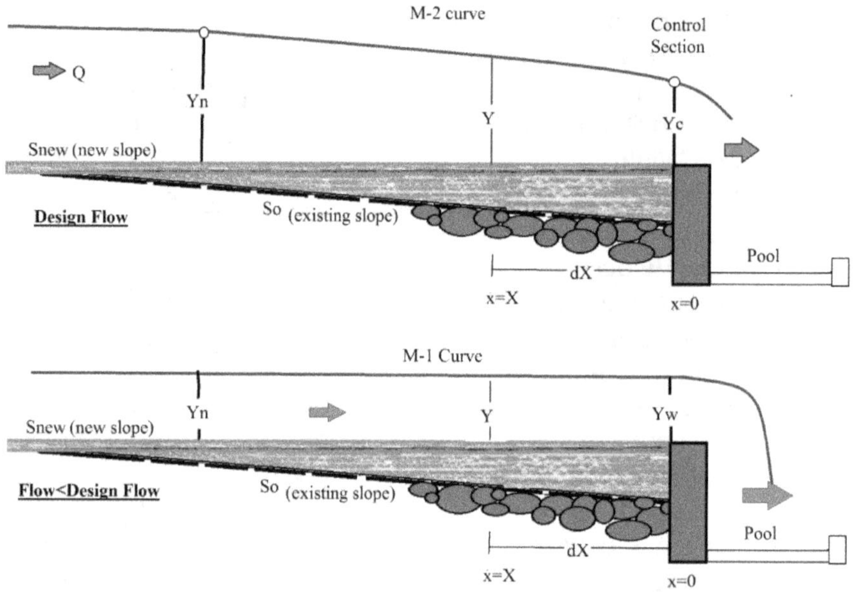

Figure 4.10. Drawdown and Backup Water Surface Profiles.

Both M-1 and M-2 water surface profiles are a steady gradually varied flow which can be computed by the principle of energy. The slope of the energy grade line (EGL) is derived by differentiating the Bernoulli's sum with respect to distance, x, as:

$$\frac{dE}{dx} = \frac{dZ}{dx} + \frac{dY}{dx} + \frac{d}{dx}\left(\frac{\alpha U^2}{2g}\right) \tag{4.5}$$

in which x = distance in [L] along channel in the flow direction. By definition, the energy slope, S_e, and channel slope, S_o, are:

$$S_e = \frac{dE}{dx} = \frac{N^2 U^2}{{k_N}^2 R^{4/3}} \tag{4.6}$$

$$S_0 = -\frac{dZ}{dx} \tag{4.7}$$

Aided by Eq's 4.6 and 4.7, Eq 4.5 is converted to:

$$\frac{dY}{dx} = \frac{S_e - S_o}{1 - F_r^2} \tag{4.8}$$

Equation 4.8 is the governing equation for gradually varied flows. The term, dY/dx, represents the variation of water surface profile. When dY/dx<0, it is *a* drawdown curve. When dY/dx>0, it is a backwater curve. And when dY/dx=0, it is a normal flow. Numerically, a finite difference scheme can be employed to solve Eq 4.8 between two adjacent sections, Y_1 and Y_2, as:

$$\frac{Y_2 - Y_1}{\Delta X} = \frac{\overline{S_e} - S_o}{1 - F_r^2} \tag{4.9}$$

$$\overline{S_e} = \frac{S_{e1} + S_{e2}}{2} = \frac{N^2}{2k_N^2}\left(\frac{U_1^2}{R_1^{4/3}} + \frac{U_2^2}{R_2^{4/3}}\right) \tag{4.10}$$

In which $\overline{S_e}$ = average energy slope in [L/L]. The subscription "1" represents the variables related to Section 1. Equation 4.9 provides the distance between the two depths.

Example 4.4: Continue with Example 4.3. After the drop is built, the ground slope is reduced from 0.02 to 0.0043. With B= 5 feet, Z=4, S_o= 0.0043 and N=0.035, the normal flow depth is determined to be 4.40 feet. The critical flow depth on top of the drop is found to be 3.39 feet. Determine the length of the M-2 curve.

Solution:

Computations for the normal flow:

Width ft B	Side Slope ft/ft Z1	Side Slope ft/ft Z2	Manning Rough N	Bottom Slope ft/ft So	Design Flow cfs Q	Normal Depth ft Yn	Wetted Prmter Pn	Flow Area ft^2 An	Hy Radius ft R	Froude No Fr	Diff in flow cfs dQ
5.00	4.00	4.00	0.035	0.0043	500.00	4.40	41.32	99.60	2.41	0.56	0.00

Computations for the critical flow:

Critical Depth Yc ft	Top Width ft Tc	Flow Area Ac	Froude Number Fr
3.39	32.13	62.95	1.00

Set $Y_1 = Y_c = 3.39$ feet on top of the drop, and $Y_2 \approx Y_n = 4.39$ feet asymptotically approaching the normal flow. The distance between the critical and normal depths is computed as:

1	2	3	4	5	6	7	8	9	10	11
Flow Depth	Flow Area	Wetted P-meter	Hy-radius	Flow Velocity	Velocity Head	Spec. Energy	ΔEs	Friciton Slope	Avg. Slope	
Y	A	P	R	U	U^2/2g	Es		Sf		X
ft	ft^2	ft	ft	fps	ft	ft	ft	ft/ft	ft/ft	ft
3.39	62.95	32.96	1.91	7.94	0.98	4.37	0.00	0.01470		
4.39	99.08	41.21	2.40	5.05	0.40	4.79	0.416	0.00436	0.00953	79.50

For this case, the distance is found to be 79.5 feet. The upstream end of the M-2 curve is asymptotic to the normal depth, to avoid a singularity (dividing by zero), the M-2 curve starts at 4.39 feet.

4.6. CLOSING

In urban area, it is a common practice to implement drops across waterways. Care must be taken if a drop is accessible by the public. The flow overtopping a drop involves a change of flow regime. High turbulent flows are expected around a drop. In comparison, a vertical drop in Figure 4.11 will produce a roller immediately downstream of the drop. It imposes a tremendous hazard to someone trapped in the roller. On the other hand, a sloping drop produces a roller immediately upstream pool of the drop. It is recommended that a sloping drop be preferred for safety.

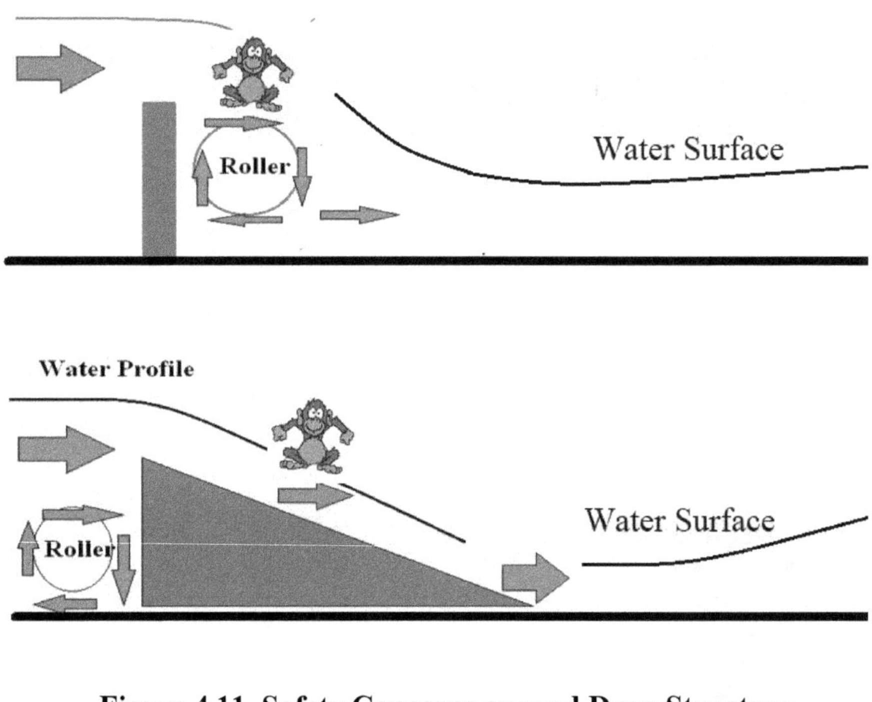

Figure 4.11. Safety Concerns around Drop Structure.

4.7. HOMEWORK

Q4.1: A trapezoidal channel has the following cross section parameters: B=5 ft, Z=4, So=0.03, and N=0.035. To reduce the flow Froude number to 0.6, determine the drop height for a reach of 100 feet.

Q4.2: A trapezoidal channel has the following cross section parameters: B=5 ft, Z=4, So=0.03, and N=0.035. To reduce the normal flow velocity to 5 fps, determine the drop height for a reach of 100 feet.

Q4.3: A trapezoidal channel in Figure HW 4.3 has a bottom width of 5 ft, side slopes of 1V:4H and 1V:3H and Manning's coefficient of 0.033. The length of channel is 200 ft on a ground slope of 0.03 ft/ft. The design discharge is 1000 cfs. Knowing that the design criteria require the flow velocity not to exceed 7 fps and the flow Froude Number not to exceed 0.80, whichever produce a higher drop height.

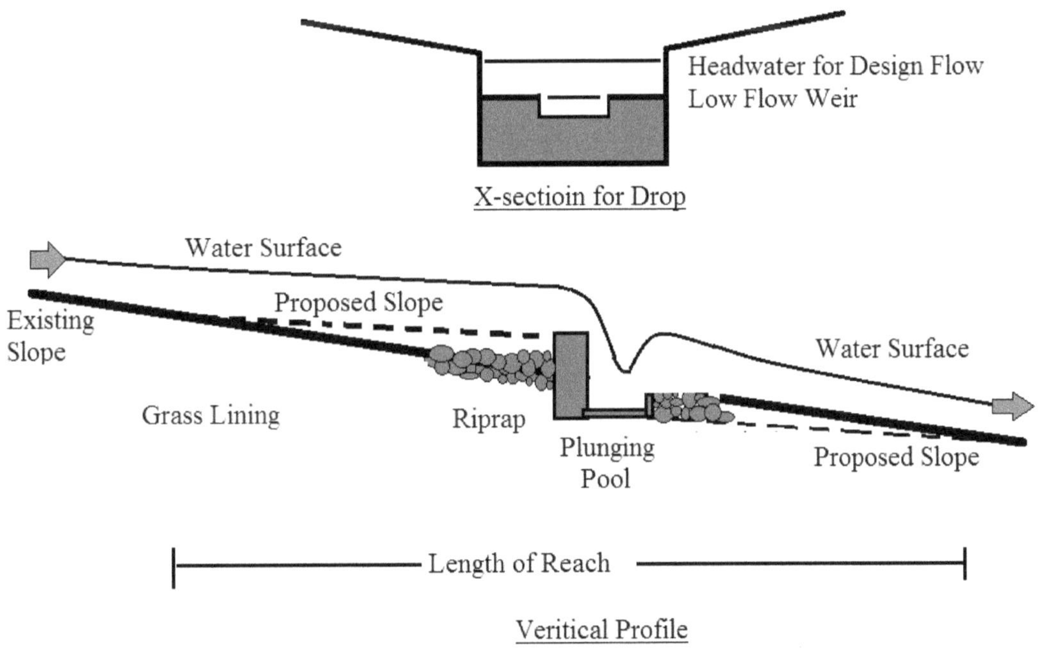

Figure HW4.3. Drop Structure Design.

(1) estimate the drop height,
(2) size the low flow weir to pass 20 cfs on top of the proposed drop structure,
(3) determine the width and water depth on top of the drop,
(4) design the plunging pool downstream of the drop,
(5) suggest the riprap rock size, D50, to protect upstream of the drop, and
(6) determine the Manning's N for Type C grass linings for the rest of the reach.

CHAPTER 5
CONCRETE CHANNEL

On a high gradient, concrete or grouted riprap lining is a good choice to withstand the high erosion potential. As a rigid wall, the side slope in a concrete channel can be as steep as 1V:1H to vertical. Thus, concrete channels as shown in Figure 5.1 are the solutions for the cases involving a narrow right of way and a high potential of scours and erosion on the stream bed. Concrete linings present the lowest surface roughness coefficient. It implies that a concrete channel provides the highest conveyance capacity to pass flood flows.

(a) Trapezoidal Concrete Channel | (b) Rectangular Concrete Channel

Figure 5.1. Concrete Channel.

5.1. DESIGN CONCERNS IN CONCRETE CHANNEL

As expected, a concrete channel is likely designed to carry shallow supercritical flows. The design criteria for concrete channel are essentially developed to control the water flow velocity for safety.

(1) Channel Cross Section

The side slope of a concrete channel can be as steep as vertical to 1V:1H. The depth-width ratio shall be determined by maximizing the hydraulic efficiency. A wide concrete channel should be tailored to have a trickle flow section to carry the low flows. A trickle channel shall have the capacity to deliver up to 1 to 3% of the major flood flow (100-yr event). In general, a trickle channel is laid through the centerline between the banks with a depth of 1 to 2 feet and width of 2 to 5 feet, depending on the magnitude of frequent flow. As demonstrated in Figure 5.2, a trickle channel confines frequent flows and significantly reduces the regular maintenance on sediment removal.

(a) Centered Trickle Channel	(b) Water-Sediment Spread

Figure 5.2. Trickle Channel.

(2) Surface Roughness Coefficient

As suggested in Table 8.1, Manning's N associated with concrete linings depends on the surface finish as:

Table 5.1. Manning's Roughness for Concrete Surface.

Surface Finish	Manning N
Trowel Finish	0.0130
Float Finish	0.0150
Unfinished	0.0170
Shotcrete, Troweled, but not Wavy	0.0180
Shotcrete, Troweled and Wavy	0.020
Shotcrete, Unfinished	0.0220

(3) Height of Freeboard

In a normal flow, the specific energy remains constant and can be interchangeable between the kinetic energy and the flow depth. Wherever the flow slows down, the water depth will sharply be raised in the concrete channel. Therefore, a freeboard of 3 feet or higher is recommended to accommodate the potential jumps and drops in a concrete channel.

(4) Strength of Concrete Panel

Often prefabricated concrete panels are used to build the channel section. The concrete panel shall have a minimum thickness of 6 inches.

(5) Permissible Flow Velocities

It is preferred that the flow velocity is between 2 and 18 ft/sec. The minimum flow velocity is capable of self-cleaning while the maximum allowable flow velocity will reduce excessive momentum forces at bends.

(6) Potential for Hydraulic Jump

Hydraulic jump is triggered in a supercritical flow if the downstream tailwater is available. Anywhere the channel is curved, a series of oblique jumps will be continuously triggered. It is advisable that all jump potentials along the channel be evaluated and the channel be protected with adequate freeboard.

(7) Superelevation of Water Surface

As shown in Figure 5.3(*a*), the curved alignment in a concrete channel results in a significant superelevation along the outer bank. An additional freeboard as shown in Fig 5.3(b) is required to avoid water spills.

(a) Superelevation	(b) Additional Freeboard

Figure 5.3. Superelevation Along Outer Bank in Curved Channel.

(8) Weeping Hole and Underdrain System

Concrete surfaces were built with prefabricated panels. Between panels exists a joint. When the seepage flows are developed through joints and cracks, a high hydrostatic pressure will be built up in the saturated soils. As shown in Figure 5.4(a), a concrete channel shall be equipped with a subdrain system that has weeping holes to release the seepage water from the saturated soil layers behind and beneath the concrete panels. Without weeping holes, as shown in Figure 5.4(b), the hydrostatic pressure from the highly saturated soils behind the grouted rock walls result in pipe failures on the bank walls.

(a) Weeping Hole	(b) Piping Failure on Bank

Figure 5.4. Weeping Hole on Concrete Bank.

Failures in concrete channels are often directly related to the high hydrostatic pressure from the saturated soil. During the recession, the flood flow is emptied out of the channel much faster than the seepage flow in soils. As shown in Figure 5.5, the unbalanced forces will collapse the concrete panels.

(a) Failure of Concrete Channel	(b) Cracks on Concrete Panels

Figure 5.5. Failures in Concrete Channels.

(9) Safety and Maintenance Access

Often a flood channel runs through residential divisions, commercial areas, or parks. Safety is a major concern in channel design. Fences are a good measure to prevent the public from approaching the hydraulic structure on a channel. Steel ladders shall be installed on the channel bank walls. Urban flood channels demand regular and frequent removals of sediment deposits within the banks. It is necessary that access ramps be installed along the channel to allow the trucks and machines to be operated for regular maintenance.

(a) Steel ladder on bank — (b) Ramp and gate for maintenance

Figure 5.6. Safety and Maintenance for Concrete Channels.

5.2. HEIGHT OF FREEBOARD

The normal flow as described in Manning's formula rarely exists. The design procedure for a new channel is to refine the normal flow depth with a freeboard The *height of freeboard* in Figure 5.7 is the vertical distance above the design water surface to the top of the banks. Safety is the main reason to have a freeboard section. With a freeboard, the channel excavated depth is equal to:

$$D = Y_n + F \tag{5.1}$$

in which D = channel depth in [L], F= height of freeboard in [L], and Y_n = normal flow depth in [L] determined by Manning's equation.

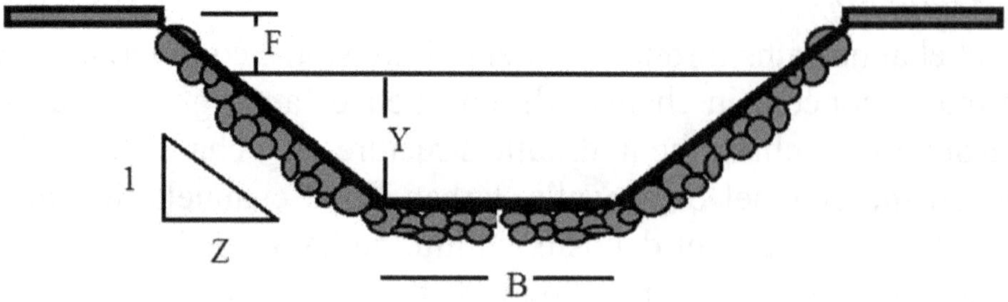

Figure 5.7. Channel Excavated Area.

The height of freeboard is often determined for the design event. Using the flow Froude number as the basis, two empirical formulas are recommended for straight reaches as:

$$F = F_B + \frac{U_n^2}{2g} \quad \text{for subcritical flow in a straight reach} \qquad (5.2)$$

$$F = 1 \text{ foot} \quad \text{if the velocity head is negligible} \qquad (5.3)$$

in which F= height of freeboard for straight reach in [L], F_B= 1.0 feet for feet-second units or 0.3 meter for meter-second units, U_n= normal flow velocity in [L/T] for the design discharge, and g= gravitational acceleration in [L/T^2]. For a supercritical flow, the height of freeboard is recommended as:

$$F = \max[\frac{1}{6}(Y_n + \frac{U_n^2}{2g}), F_H] \quad \text{for supercritical flow in a straight reach} \qquad (5.4)$$

in which F_H= 3.0 feet for feet-second units or 1.0 meter for meter-second units. Caution must be taken when coping with a supercritical flow. For instance, the kinetic energy for a flow velocity of 30 fps can be converted into a water depth of approximately 14 feet. It implies that the flow has a potential to jump to 14 feet under a blocking condition. Therefore, it is important to intall a sufficient freeboard section on a high gradient channel.

5.3. SUPERELEVATION

As illustrated in Figure 5.8, the water rise around the outer bank along a curved channel is termed superelevation. Based on flow dynamics, the centrifugal force supporting the curve motion is translated into the rise of water depth as:

$$h = \frac{U_n^2 T_n}{g R_c} \qquad (5.5)$$

in which h = superelevation in [L] in curve reach, T_n = top width in channel in [L], and R_c = centerline radius of curvature in [L]. At a curved reach, the height of freeboard is determined by the higher between Eq 5.5 and Eq 5.2 or 5.3.

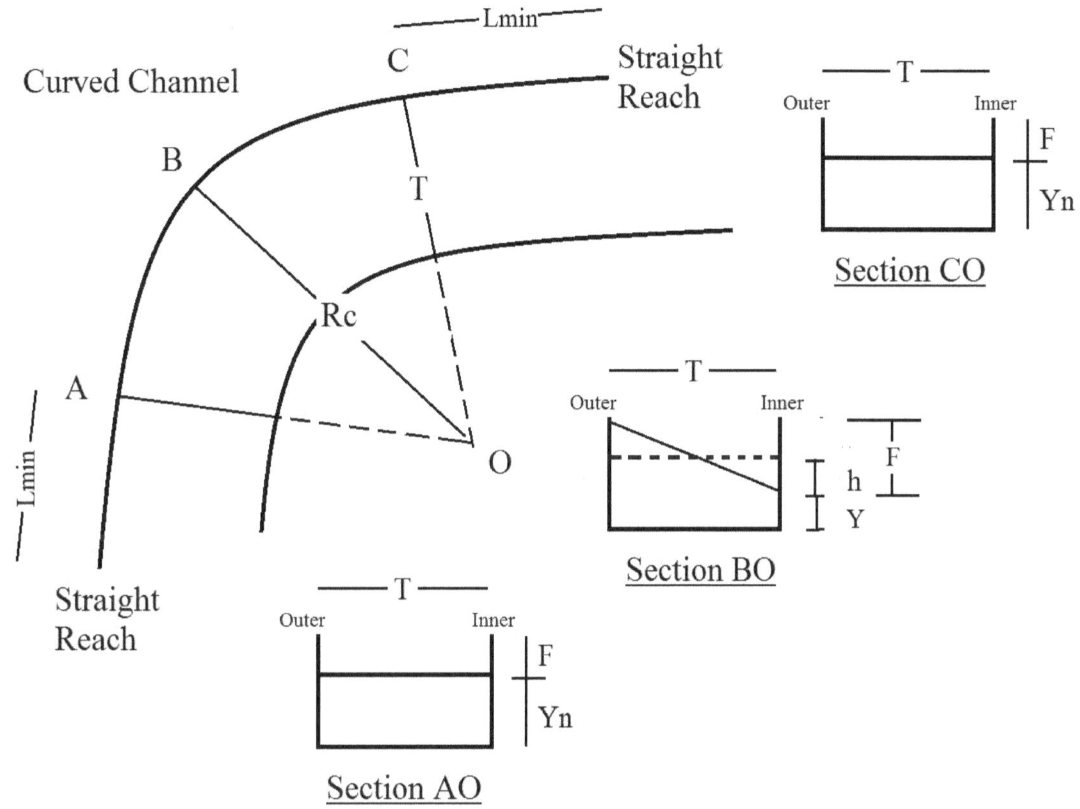

Figure 5.8. Transition from Straight to Curved Reaches.

It is suggested that the radius of curved alignment be at least three times the top width of the channel as:

$$R_c \geq 3T \tag{5.6}$$

The superelevation increases the height of freeboard. As expected, there is a transition from the incoming straight reach into the curve reach, so is from the curve reach into the outgoing straight reach. The minimum length, L_{min} in [L], for the transitional reach is recommended to be:

$$L_{\min} = 30h \tag{5.7}$$

In urban areas, a narrow right of way presents a challenge to design curve channels. Of course, a closed culvert system may be a substitute when the site is limited for an adequate radius of curvature.

Example 5.1: A trapezoidal channel is designed to carry 1000 cfs with B= 2 feet, Z=1, S_o= 2.0%. The concrete panels are prefabricated with an unfinished shotcrete surface. Design the curved reach for this channel.

Solution: As recommended in Table 5.1, N=0.022 for this channel. The normal flow condition for B=2 feet, Z=0, and S_o=0.02 is determined as:

Flow Depth	Top Width	Flow Area	Wetted P-meter	ft Radius	Flow Velocity	Flow rate	Froude Number	Sp Energy	Depth to Centroid	Sp Force
Y	T	A	P	R	U	Q	Fr	Es	Yh	Fs
ft	ft	sq ft	ft	ft	fps	cfs		ft	ft	klb
6.42	14.84	54.1	20.2	2.68	18.50	1000.0	1.71	11.72	2.38	43.91

This channel will have a flow Froude number of 1.71>1.0. Therefore, the height of freeboard, radius of curvature, and transitional approaching length are:

$$F = \max[\frac{1}{6}(Y_n + \frac{U_n^2}{2g}),\ F_H] = \max(\frac{11.72}{6}, 3.0) = 3.0 \ \textit{feet}$$

$R_c \geq 3 \times 14.84$ *feet.* *Try R_c= 45 feet.*

$$h = \frac{U_n^2 T_n}{gR_c} = \frac{18.5^2 \times 14.84}{32.2 \times 45} = 3.52 \text{ feet}$$

$$L_{\min} = 30h = 105 \ \textit{feet}$$

For this case, the height of freeboard is determined by the higher between the empirical formula for straight reach and the superelevation for curve reach as:

F=max(3.0, 3.27)=3.27 feet.

5.4. EFFICIENT CHANNEL SECTIONS

As indicated in Manning's formula, the hydraulic capacity in a rigid channel solely depends on the sectional elements and the longitudinal slope. For convenience, Manning's formula is further simplified to:

$$Q = K\sqrt{S_0} \tag{5.8}$$

in which Q= design discharge in [L^3/T], S_o= channel slope in [L/L], and K = conveyance of channel section defined as:

$$K = \frac{k_N}{N} A_f^{\frac{5}{3}} P^{\frac{-2}{3}} \tag{5.9}$$

where k_N=1.486 for feet-second units or 1.0 for meter-second, N= Manning's roughness coefficient, A_f= flow area in [L^2], and P=wetted perimeter in [L]. As shown in Figure 5.7, the excavated area of a channel section includes both the flow area and the required freeboard height, F. For a trapezoidal channel, we have

$$A_t = (Y+F)B + Z(Y+F)^2 \tag{5.10}$$

$$A_f = YB + ZY^2 \tag{5.11}$$

in which A_t = excavated area in [L^2], Y = flow depth in [L], B = bottom width in [L], Z = side slope in [L/L], and F = height of freeboard in [L]. From a cost aspect, the optimal channel section is defined to be the cross section that provides the minimum excavation for the given discharge. Namely:

Objective Function = Minimization of cost= Min excavated area, A_t
subject to: Q *= given design discharge*

Notice that the frictional resistance to water flow is proportional to the wetted perimeter. Therefore the minimization of the excavated area is equivalent to the minimization of hydraulic resistance. As a result, the above objective will lead to the most hydraulically efficient channel section as:

Objective Function = Minimization of friction= Maximization of capacity, Q
subject to: A_t *= specified constant*

As a result, it is concluded that the optimal channel section can be achieved by either maximizing the capacity for a specified channel sectional area or minimizing the excavated area for a specified capacity. A specified channel sectional area can be translated into a fixed cost while a specified capacity implies a given design discharge. Both objective functions discussed above lead to the same conclusion as (Guo and Hughes 1984):

$$\frac{dA_t}{dY} = 0 \quad \text{(Minimization)} \tag{5.12}$$

And

$$\frac{dQ}{dY} = \frac{dK}{dY} = 0 \quad \text{(Maximization)} \tag{5.13}$$

In general, the value of Z is pre-determined by the soil stability such as the angle of repose. As a result, K in Eq 5.9 and A_t in Eq. 5.10 and are dependent of both depth Y and bottom width B. Therefore, the total derivative of A_t is written as:

$$\frac{dA_t}{dY} = \frac{\partial A_t}{\partial Y} + \frac{\partial A_t}{\partial B}\frac{dB}{dY} = 0 \tag{5.14}$$

Similarly, the total derivative for Eq. 5.13 is:

$$\frac{dK}{dY} = \frac{\partial K}{\partial Y} + \frac{\partial K}{\partial B}\frac{dB}{dY} = 0 \tag{5.15}$$

The optimal trapezoidal section is expressed by its depth-to-width ratio. Substituting Eq 5.10 into Eq 5.14 yields:

$$\frac{dB}{dY} = -\frac{B}{Y+F} - 2Z \tag{5.16}$$

For convenience, let the following variables be defined as:

$$m = \frac{F}{Y} \tag{5.17}$$

$$n = \frac{B}{Y} \tag{5.18}$$

$$Q' = \left(\frac{NQ}{k_N \sqrt{S_0}} \right)^3 \tag{5.19}$$

The variable m is the ratio of freeboard height to flow depth, and n is the ratio of channel bottom width to flow depth. Substituting Eq's 5.17 through 5.19 into Eq 5.16 yields the optimal trapezoidal shape to be:

$$n = \frac{1}{5m+2}\left[-(3km + 2Zm - 2k + 3Z) + \sqrt{M} \right] \tag{5.20}$$

$$M = 9k^2m^2 - 16Z^2m^2 + 4k^2 + Z^2 + 32m^2kZ - 12k^2m + 38mkZ - 16Z^2m - 4kZ \tag{5.21}$$

$$k = \sqrt{1+Z^2} \tag{5.22}$$

$$A'_t = \frac{A_t}{Y^2} = n(1+m) + Z(1+m)^2 \tag{5.23}$$

$$A'_f = \frac{A_f}{Y^2} = n + 2 \tag{5.24}$$

$$P' = \frac{P}{Y} = n + 2k \tag{5.25}$$

$$Q'' = \frac{Q'}{Y^8} = (n+Z)^5 (n+2k)^{-2} \tag{5.26}$$

Several special cases need to be further discussed.

5.4.1. Efficient Trapezoidal Channel without Freeboard

Without a freeboard, we have m = 0. Equation 5.20 is reduced to:

$$n = \frac{B}{Y} = 2(k - Z) \tag{5.27}$$

Substituting Eq 5.27 into Eq's 5.23 through 5.26 yields

$$\frac{A_t}{Y^2} = n + Z \tag{5.28}$$

$$\frac{A_f}{Y^2} = 2k - Z \tag{5.29}$$

$$\frac{P}{Y} = 4k - 2Z \tag{5.30}$$

$$\frac{Q'}{Y^8} = (2k - Z)^5 (4k - 2Z)^{-2} \tag{5.31}$$

To minimize the wetted perimeter, we can differentiate Eq 5.30 with respect to Z and then set it equal to zero. The best side slope is found to be:

$$Z = \frac{1}{Z_0} \tag{5.32}$$

in which $Z_o = \sqrt{3}$. Equation 5.32 implies that the most efficient trapezoidal channel section is a half hexagon. Substituting Eq 5.32 into Eq's 5.26 through 5.31 yields

$$\frac{A_t}{Y^2} = Z_0 \tag{5.33}$$

$$\frac{A_f}{Y^2} = Z_0 \tag{5.34}$$

$$\frac{P}{Y} = 2Z_0 \tag{5.35}$$

$$\frac{Q'}{Y^8} = \frac{Z_0^{\,3}}{4} \tag{5.36}$$

5.4.2. Efficient Rectangular Channel with Freeboard

Since $Z = 0$ and $k = 1$, Eq 5.20 becomes:

$$n = \frac{B}{Y} = \frac{2(2 - 3m)}{(5m + 2)} \tag{5.37}$$

Substituting Eq 5.37 into Eq's 5.23 through 5.26 yields:

$$\frac{A_t}{Y^2} = \frac{2(2-3m)(1+m)}{(5m+2)} \tag{5.38}$$

$$\frac{A_f}{Y^2} = \frac{2(2-3m)}{(5m+2)} \tag{5.39}$$

$$\frac{P}{Y} = \frac{4(2+m)}{(5m+2)} \tag{5.40}$$

$$\frac{Q'}{Y^8} = \frac{2(2-3m)^5(2+m)^{-2}}{(5m+2)^3} \tag{5.41}$$

5.4.3. Efficient Rectangular Channel with No Freeboard

For a case of rectangular channel without a freeboard, we have $F = 0$, $Z = 0$, $m = 0$, and $k = 1$, Eq 5.37 becomes

$$n = \frac{B}{Y} = 2.0 \tag{5.42}$$

Substituting Eq 5.42 into Eq's 5.36 through 5.41 yields

$$\frac{A_t}{Y^2} = 2.0 \tag{5.43}$$

$$\frac{A_f}{Y^2} = 2.0 \tag{5.44}$$

$$\frac{P}{Y} = 4.0 \tag{5.45}$$

$$\frac{Q}{Y^8} = 2.0 \tag{5.46}$$

Equations 5.43 through 5.46 agree with the conventional solution for the optimal rectangular channel without a freeboard (Chow 1959).

Example 5.2: Continue Example 5.1. With Q= 1000 cfs, Z=1, B=2 feet, N=0.022, determine the optimal channel section with a freeboard of 3.27 feet.

Solution: The following iterative procedure is recommended to derive the solution for the optimal channel cross section:

(1) Guess the value of m. Set m=0.51

(2) Compute

$$Q' = \left(\frac{NQ}{k_N \sqrt{S_0}} \right)^3 = 1138058 \text{ where } k_N = 1.486$$

(3) Determine the ratio Q'/Y^8 by Eq 5.41 as:

$$\frac{Q'}{Y^8} = \frac{2(2-3m)^5 (2+m)^{-2}}{(5m+2)^3} = 0.38. \text{ So, } Y = 6.43 \textit{ feet}$$

(4) Calculate the value of *m*.

$$m = \frac{F}{Y} = \frac{3.27}{6.43} = 0.508 \approx 0.51$$

The computed value of m agrees with the guessed value of 0.51. If not, repeat the dame process with another value of m.

(5) Substituting m=0.54 into Eq's 5.38 to 5.40 produce:

$n = \frac{B}{Y} = 0.31$. So, B=1.98 feet (Noticed that B=2 ft was used in Example 5.1)

$$\frac{A_t}{Y^2} = 2.75. \quad \text{So, } A_t = 113.51 \text{ ft}^2$$

$$\frac{A_f}{Y^2} = 1.31, \quad \text{So, } A_f = 54.1 \text{ ft}^2$$

$$\frac{P}{Y} = 3.14, \qquad \text{So,} \;\; P = 20.17 \;\; \text{ft}$$

For this case, a sensitivity study on Z is further conducted for Z=1 to 7. As an iterative process, the computations shall begin with the guessed ratio: m=h/Y. For this case, the value, m=0.54, generates the optimal solution with F= 3.27 feet and B= 1.98 (≈2 feet) which are close to the design parameters and requirements.

Freeboard m=h/Y = 0.510

Side Slope	Side Parameter	M-Parameter	Width to Depth Ratio	Total Area Ratio	Flow Area Ratio	Wetted P-meter Ratio	Flowrate to Depth Ratio
Z	$k=(1+z^2)^{.5}$	M	n=B/Y	A_t/Y^2	A_f/Y^2	P/Y	Q'/Y^8
1.00	1.41	22.64	0.31	2.75	1.31	3.14	0.39
2.00	2.24	61.82	0.19	7.00	2.19	4.66	2.33
3.00	3.16	125.18	0.14	7.04	3.14	6.46	7.26
4.00	4.12	213.53	0.10	9.28	4.10	8.35	16.69
5.00	5.10	327.02	0.08	11.53	5.08	10.28	32.12
6.00	6.08	465.68	0.07	13.79	6.07	12.24	55.04
7.00	7.07	629.54	0.06	16.05	7.06	14.20	86.97

Side Slope	Flow Depth	Bottom Width	Free Board	Wetted Parameter	Flow Area	Excavated Area	Froude Number	Specific Energy
Z	*Y*	*B*	*F*	*P*	*Af*	*At*	*Fr*	*Es*
	feet	feet	feet	feet	sq ft	sq ft	feet	feet
1.00	6.43	1.98	3.28	20.17	54.09	113.51	1.71	11.74
2.00	5.14	0.99	2.62	23.99	57.97	185.13	1.85	9.76
3.00	4.46	0.60	2.28	28.82	62.39	140.17	1.87	8.45
4.00	4.02	0.42	2.05	33.57	66.31	149.91	1.86	7.55
5.00	3.70	0.31	1.89	38.08	69.75	158.15	1.85	6.90
6.00	3.46	0.24	1.77	42.37	72.79	165.32	1.83	6.39
7.00	3.27	0.20	1.67	46.45	75.51	171.68	1.82	5.99

It is also noticed that the higher the side slope, Z, the shorter the width B. It implies that the optimal trapezoidal cross section is asymptotically to triangular shape.

5.5. CLOSING

The following is the check list for concrete channel design. The engineer shall check out each item on the list to make sure the design task is completed.

Table 5.2. Design Submittal Checklist for Concrete-Lined Channel.

Criterion/Requirement	✓
Maximum velocity for 100-year event ≤ 18 ft/sec	
Channel capacity checked with Manning's $n = 0.013$	
Velocity checked using Manning's $n = 0.011$	
Froude number ≤ 0.7 based on capacity check or ≥ 1.4 based on velocity check	
Side slopes no steeper than 1.5H:1V	
Centerline curve radius > 2 x top width for 100-year event and not < 100 ft for subcritical channels	
No curvature for supercritical channels	
Concrete lining designed to withstand hydrodynamic and hydrostatic forces (minimum thickness = 7.0 inches for supercritical channels, 5.0 inches for subcritical channels)	
Concrete joints meet Section 4.3.2.2 criteria	
Free draining granular bedding under concrete (minimum thickness = 6 inches for Fr ≤ 0.7 and 9 inches for Fr > 0.7	
Concrete finish from list in Table MD-5	
Outfalls into channel ≥ 1 ft above channel invert	
Adequate freeboard provided (see criteria in text)	
Standing waves included in freeboard for supercritical channels	
Free draining longitudinal underdrains provided on 10-ft centers, including check or flap valve at outlet to prevent backflow	
6-ft chain link fence (or comparable) provided when channel depth > 3.0 ft	
Ladder-type s spaced no more than 200 ft apart on alternating sides of channel with lowest rung approximately 12 inches above channel invert	
District Maintenance Eligibility Guidelines satisfied	
Continuous maintenance access road provided (minimum 8-foot stable surface with 12-foot clear width)	
Energy and hydraulic grade lines calculated and plotted for channel	

5.6. HOMEWORK

Q5.1: The concrete rectangular channel is designed to carry a flow of 100 cfs on a slope of 1%. Consider N=0.014, Z=0, F=0, find the optimal water depth.

Solution: The optimal rectangular section is to have:

$B = 2Y,\ A = 2Y^2,\ and\ P = B + 2Y = 4Y.$

$$Q = 100 = \frac{1.486}{0.014} A^{5/3} P^{-2/3} \sqrt{0.01} = 10.6(2Y^2)^{5/3}(4Y)^{-2/3}$$

So, Y=2.12 ft

Q5.2: The concrete trapezoidal channel in Example 5.1 carries a flow of 1000 cfs on a slope of 2%.

(1) with F=0 and Z=0, determine the optimal cross section

(2) with F=0 and Z=1, determine the optimal cross section

(3) with F=1.0 ft and Z=0, determine the optimal cross section, and

(4) with F=1.0 ft and Z=1, determine the optimal cross section.

Solution:

Side Slope	Side Parameter	M-Parameter	Width to Depth Ratio	Flow Depth	Bottom Width	Free Board
z	k=(1+z2).5	M	n=B/Y	Y	B	F
				feet	feet	feet
0.00	1.00	4.00	2.00	4.42	8.85	0.00
1.00	1.41	3.34	0.83	4.57	3.79	0.00

Side Slope	Side Parameter	M-Parameter	Width to Depth Ratio	Flow Depth	Bottom Width	Free Board
z	k=(1+z2).5	M	n=B/Y	Y	B	F
				feet	feet	feet
0.00	1.00	2.28	1.07	6.11	6.56	1.00
1.00	1.41	7.85	0.47	5.11	2.41	1.00

CHAPTER 6
GRASS CHANNEL

Grass lining in Figure 6.1 is to mimic natural channels. Hydraulic properties of a natural channel are varied along the channel reach because the cross sections are irregular. The floodplains along a natural channel have been shaped by long term erosions and scours by flood flows. After a natural channel reaches the dynamic equilibrium, the sediment transport and deposit are in balance. However, urban development increases runoff rates and volumes continually. As a result, the morphology of an urban flood channel is subject to urbanization impacts, and the streambed is unstable due to degradation or aggregation.

Straight Natural Channel	(b)Meandered Natural Channel

Figure 6.1. Natural Channel.

To stabilize a natural channel, the engineer must prepare the channel cross sections and collect field data such as sediment samples, bank materials etc. Studies of the existing floodplain condition provide the basic understanding of the channel morphology and serve as a basis to evaluate the impacts of the urban development on the natural channel. The recommended criteria and considerations for natural channel improvements are:

(1) Natural channels shall have adequate capacity to pass the major storm event,

(2) Floodplains of a major drainage way must be defined and published,

(3) Roughness coefficients must be carefully determined,

(4) Velocity and Froude number for the design discharge must be computed,

(5) Any erosion control structures must be protected from erosion.

The design procedure for grass channels is to observe natural channels and follow the concept of permissible velocity to stabilize the proposed cross sections within the right of way.

6.1. GRASS LININGS

Grass-linings are the most economic and desirable for urban channels. Grass roots provide good protection to the stream bed from erosion and embankments from scours. As shown in Figure 6.2, the vegetal resistance is closely related the height of grass stems. The hydraulic resistance of various types of grass linings has been classified into five categories.

(1) Class A : *very high vegetal retardance,*

(2) Class B : *high vegetal retardance*

(3) Class C : average *vegetal retardance*

(4) Class D : *low vegetal retardance*

(5) Class E : *very low vegetal retardance*

(a) Type A Grass Lining

(b) Type B/C Grass Channel

(c) Type D/E Grass Channel

(d) Landscaping Grass Channel

Figure 6.2. Grass Linings.

Table 6.1 provides the classification of hydraulic resistance for various grass linings.

Table 6.1. Retardance of Grass Linings.

Retardance	Cover	Condition	Height in Inches
A	Weeping Lovegrass	Excellent Stand	30 to 36
B	Kudzu Bermudagrass Weeping Lovegrass Alfalfa Mixed Grass	Dense Growth and Uncut Good Stand Good Stand Good Stand Uncut and Unmowed	 12 24 12
C	Crabgrass Bermudagrass Common Lespedeza Mixed Grass Centipedegrass Kentucky Blue	Fair Stand Good Stand, Mowed Uncut Good Stand, Uncut Very Dense Good Stand	10 to 48 6 11 6 to 8 6 11
D	Bermudagrass Common Lespedeza Buffalograss Mixed Grass Lespedeza	Good Stand, Cut Excellent Stand, Uncut Good Stand, Uncut Good Stand, Uncut Good Stand, Cut	2.5 4.5 3 to 6 4 to 5 2
E	Bermudagrass	Good Stand, Cut	1.0 to 1.5

6.2. MANNING'S ROUGHNESS FOR GRASS LININGS

The roughness coefficient, Manning's N, in a grass channel has been found to be a function of the product, *UR*, in which *U* is the average velocity and *R* is the hydraulic radius. As shown in Figure 6.3, the design curves developed by the US Natural Resource Conservation Services (NRCS) or the former Soil Conservation Service (SCS) are recommended for determining Manning's N by an iterative process. It is noticed that Manning's N reaches a limiting value for all types of grass linings. For instance, the Manning's N for Type A grass converges to 0.06. These minimum values for various grass linings listed in Table 6.2 may serve as an initial guessed roughness coefficient for the iterative process to design a grass lined channel.

Table 6.2. Limiting Manning's N for Grass Linings.

Type of Grass Linings	A	B	C	D	E
Minimum Manning's N	0.0600	0.0400	0.0330	0.0300	0.0240

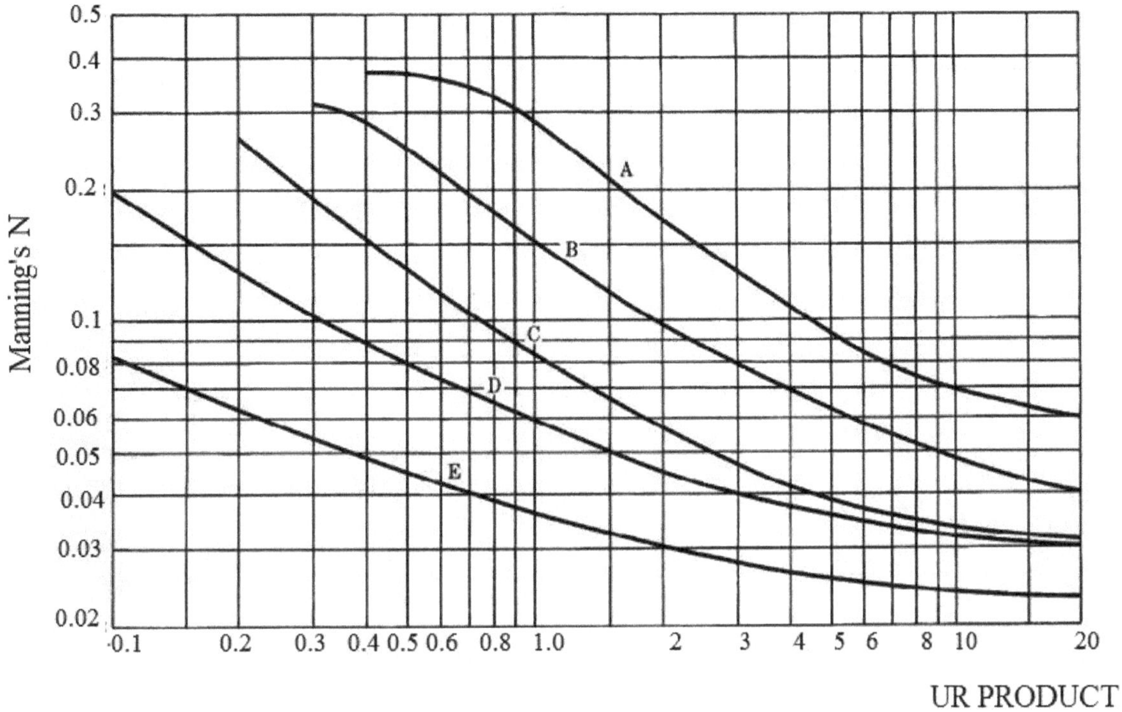

Figure 6.3. Manning's N For Grass Linings.

6.3. GRASS CHANNEL DESIGN

The major concern in the design of grass channel is the potential erosion due to flow velocities. The recommended design criteria for a grass-lined channel include:

(1) The maximum permissible velocity for the major storm event in a grass channel is 7.0 fps for cohesive soil and 5.0 fps for non-cohesive soil. A minimum of 2.0 fps should be maintained for the minor event.

(2) Froude number for the design discharge must be less than 0.7 for cohesive soil, and 0.6 for sandy soil.

(3) Where the natural topography is steeper than desired, a drop structure may be considered to reduce the channel grade.

(4) For convenience in maintenance, the flow depth for the major event shall be kept less than five feet.

(5) It is preferable to have a low-flow trickle channel through the thalweg of a grass channel. The trickle channel in Figure 6.4 shall be sized to pass 1 to 3% of the design flow. The trickle channel may be lined with concrete panel or riprap rocks.

(a) Concrete trickle channel | (b) Riprap trickle channel

Figure 6.4. Trickle Low-flow Channel in Grass Channel.

To recognize the potential scour, a grass lined channel is designed to have the maximum permissible velocity, U_m, under the design discharge not to exceed 7.0 fps for cohesive soil or 5.0 fps for sand soil.

$$A_m = \frac{Q}{U_m} \tag{6.1}$$

The design process of a grass channel is iterative until the criteria are met. It may start with an assumed Manning's N which is equal to the minimum value for the type of grass as shown in Table 6.2. The hydraulic radius corresponding to the maximum permissible velocity can then be determined as follows:

$$R = \left(\frac{U_m N}{k_N \sqrt{S_o}}\right)^{3/2} \tag{6.2}$$

Next, we shall compute the product, R.U_m, by which we can revise the Manning's N using Figure 6.3. Substituting the new N value into Eq 6.2 yields a new value for R. This iterative process may stop until the difference between the new and old Manning's N's is less than 1%.

$$Error = \left| \frac{N_{n+1} - N_n}{N_n} \right| < 0.01 \tag{6.3}$$

in which N_n = Manning's N at n-th update. Upon the convergence of Manning's N, the flow area and hydraulic radius become known by Eq's 6.1 and 6.2.

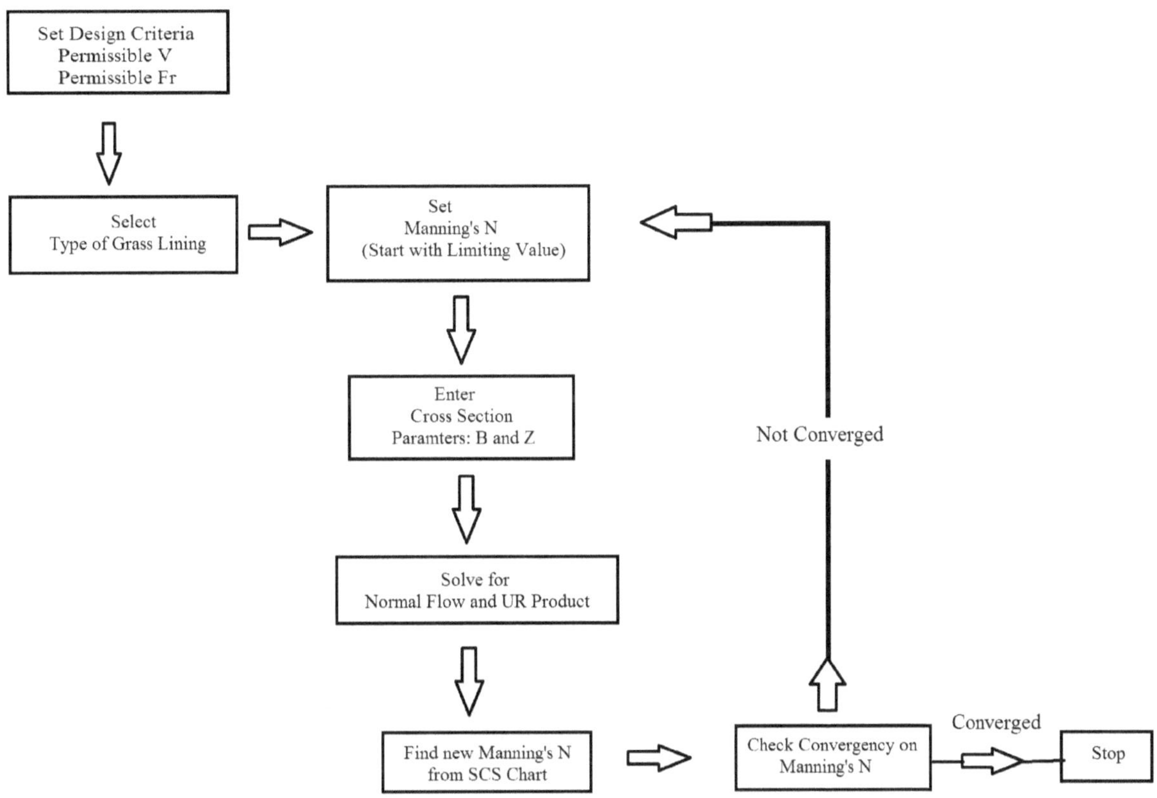

Figure 6.5. Iterative Process for Grass Channel Design.

The cross sectional elements including flow depth, Y, channel width, B, side slope, Z, can then be determined by solving the following two equations simultaneously.

$$A = \frac{Q}{U_m} \tag{6.4}$$

and

$$R = \frac{A}{P} \tag{6.5}$$

Equations 6.4 and 6.5 offer multiple solutions for the pair of (B,Z). In practice, the engineer shall pre-select the pair (B,Z) and solve for the normal flow condition. Figure 6.5 summarizes the grass channel design iteration procedures.

Example 6.1: Continue with Example 4.1 in which the channel is designed to carry a design flow of 2000 cfs in a grass-lined trapezoidal cross section with bottom width B= 15 feet and side slope Z=4.0. The existing slope is reduced from S_o=0.03 to S_o=0.0042 with a drop height of 4.64 feet per 180-feet reach. Determine the grass lining for this channel.

Solution: Try Type D grass. The limiting Manning's N=0.04 and the normal flow are computed as:

Design Guideline				
Permissible Flow Velocity	U-max=	7.00	fps	
Permissible Froude Number	Fr-max=	0.80		
Design Information				
Type of Grass (A,B,C,D, or E)		D		
Channel Invert Slope	So =	0.00420	ft/ft	
Bottom Width	B =	15.00	ft	
Left Side Slope	Z1 =	4.00	ft/ft	
Right Side Slope	Z2 =	4.00	ft/ft	
Design Flow Rate	Q=	2000.00	cfs	
Enter Grass Manning's N	N =	0.0400	**Guess**	
Calculation of Flow Depth				
Water Depth	Y =	7.21	ft	**Guess**
Top Width	T=	72.67	ft	
Flow Area	A=	316.01	sq ft	
Wetted Perimeter	P=	74.45	ft	
Hydraulic Radius	R=	4.24	ft	
Hydraulic Depth	D=	4.35	ft	
Flow Velocity	U=	6.33	fps	<Umax
Froude Number	Fr =	0.53		<Fr-max
Difference in Q	dQ=	0.00	cfs	**Check**
Check on N-UR relationship				
UR product	UR=	**26.86**		
New N from Design Chart	N-new=	**0.0400**	**Read**	
Error abs(Nnew - N)/N	% =	**0.00**		

6.4. CLOSING

Grass-lined channels are designed under the constraints of permissible flow velocity and Froude number. Often it is necessary to have drops to reduce the energy slope associated with the flow. It takes time for grass linings to become mature and strong enough to sustain the expected vegetal resistance. During the seeding stage as shown in Figure 6.6a, the channel is vulnerable to flood flows. It takes a couple of years for grass roots to grow and to become stable.

(a) Before grass seeding | (b) Seeding with covers | (c) Well maintained | (d) Reinforced grass-lining

Figure 6.6. Construction Stages of Grass Channel.

For a matured grass channel, the maintenance along a grass channel is to keep the height of grass stem the same as designed. In practice, any grass taller than 36 inches tend to be bended over to form a grass blanket during a high flood flow. A grass lining can be strengthened with "Turf Reinforcement Mats (TRMs)". TRMs is available as a three-dimensional, lofty, woven polypropylene or stitch-bonded geotextile that delivers long-term, value engineered erosion protection. Table 6.3 is the check list to complete a grass channel design.

Table 6.3. Design Submittal Checklist for Grass-Lined Channel.

	✓
Maximum velocity for 100-year event: ≤ 5.0 ft/sec for erosive soils ≤ 7.0 ft/sec for non-erosive soils	
Manning's $n \geq 0.035$ used to check capacity	
Manning's $n \leq 0.030$ used to check velocity	
Froude number: < 0.5 for erosive soils < 0.8 for non-erosive soils	
Maximum depth for 100-year event ≤ 5.0 ft outside of low-flow channel	
Longitudinal channel slope ≥0.2% and ≤ 0.6%	
Side slopes no steeper than 4H:1V	
Channel bottom cross slope ≥ 2%	
Low-flow section or outside toes of main channel extended below low-flow invert lined with VL rock mixed with topsoil, buried and vegetated	
Centerline curve radius > 2 x top width for 100-year event and not < 100 ft	
Outfalls into channel ≥ 1 ft above channel invert (use pipes, concrete-lined rundowns or grouted boulder rundowns)	
Adequate freeboard provided, including superelevation	
Grass species appropriate (drought resistant, sturdy, easily established, turf forming)	
Trickle channel (if any) sized for 2.0% of 100-year design flow for fully developed, undetained condition	
Low-flow channel (if any) sized for 1/3 to 1/2 of the 2-year design flow for fully developed condition	
Low-flow channel (if any) depth ≥ 3.0 ft and ≤ 5.0 ft	
Underdrain pipe (if any) diameter ≥ 24 inches [Note: not recommended or endorsed.]	
Underdrain pipe (if any) includes manhole access every 200 ft	
Underdrain pipe (if any) velocity ≥ 3.0 ft/sec when half-full	
Channel bends checked for need for erosion protection	
Erosion protection measures included where necessary	
District Maintenance Eligibility Guidelines satisfied	
Continuous maintenance access road provided (minimum 8-foot stable surface with 12-foot clear width)	
Energy and hydraulic grade lines calculated and plotted	

6.5. HOMEWORK

Q6.1: A trapezoidal channel has a bottom width, B=10 feet, and side slope: Z=4. The channel slope is: So=0.52%. This channel is covered with SCS Type C grass lining using 6-inch Bermuda grass. The design discharge in this channel is 1200 cfs. Verify that this channel satisfy the design criteria, including the permissible velocity <=7.0 fps and the flow Froude number =< 0.75.

Solution:

Design Guideline

Grass Type	A	B	C	D	E
Limiting Manning's N	0.060	0.040	0.033	0.030	0.024

Non-erodable flow velocity	U-max=	7.00	fps	
Limiting Froude Number	Fr-limit=	0.75		
A. Design Information				
Type of Grass (A,B,C,D, or E)		C		
Channel Invert Slope	So =	0.00520	ft/ft	
Bottom Width	B =	10.00	ft	
Left Side Slope	Z1 =	4.00	ft/ft	
Right Side Slope	Z2 =	4.00	ft/ft	
Design Flow Rate	Q=	1200.00	cfs	
B. Enter Grass Manning's N	N =	0.0330	Guess	
C. Calculation of Flow Depth				
Water Depth	Y =	5.42	ft	**Guess**
Top Width	T=	53.38	ft	
Flow Area	A=	171.85	sq ft	
Wetted Perimeter	P=	54.72	ft	
Hydraulic Radius	R=	3.14	ft	
Hydraulic Depth	D=	3.22	ft	
Flow Velocity	U=	6.98	fps	<Umax
Froude Number	Fr =	0.69		<Fr-limit
Difference in Q	dQ=	0.00	cfs	**Check**
D. Check on N-UR relationship				
UR product	UR=	**21.93**		
New N from Design Chart	N-new=	**0.0330**	Read	
(N-new - N)/N	% =	**0.00**		

Replace N with N-new and calculate UR product again.

CHAPTER 7
RIPRAP CHANNEL

Riprap linings are flexible and generally less expensive than rigid linings. Riprap lining is a good choice if the channel is expected to experience frost heave and/or swelling of the underlying soils that can change the shape of the channel lining. Riprap blankets permit infiltration and exfiltration and can be vegetated to have a natural appearance. More natural behavior offers better habitat opportunities for local flora and fauna. In many cases, riprap linings are designed to provide only transitional protection against erosion before vegetation becomes the permanent linings in the channel. Vegetative channel lining is recognized as a best management practice for storm water quality enhancement design.

7.1. RIPRAP LINING

Riprap lining in Figure 7.1 is suitable for a short but steep channel reach. The nature of high friction of rocks contributes to the effectiveness of energy dissipation. Especially where right of way for the drainage way is limited, riprap channel may be considered because steeper side slopes can be acceptable.

(a) Grouted Riprap Channel | (b) Rock Riprap Channel

Figure 7.1. Riprap Channels.

7.2. DESIGN CONSIDERATIONS

Riprap linings provide protection to the streambed from erosion and scours. Design constraints are listed as follows:

(1) the permissible velocity for riprap linings is 12.0 feet/second,
(2) the safety factor for shear stress>1.0 to 1.5.
(3) the roughness coefficient, Manning's N, is ranged from 0.035 to 0.045,
(4) the side slope for a riprap cross section can be up to 1V:2H, and
(5) the rock specific gravity is 2.5 to 2.7 for design.

Motions of particles on a streambed are closely related to the flow shear stress or the stream energy. There are two approaches recommended to design riprap lining. The *shear stress-based method* is recommended for reaches with a supercritical flow while the *stream power-based method* is recommended for subcritical flows.

7.3. STREAM POWER-BASED METHOD

The stream power-based method was developed using the product of US_0 as the basis to determine the riprap rock size, in which U is the cross sectional average flow velocity and S_0 is the channel bottom slope as a substitute for energy slope. In practice, a high gradient channel can be improved with a drop structure to reduce the energy slope and to convert the flow regime from a supercritical to subcritical flow. Often the immediately upstream reach of a drop structure shall be protected with riprap rocks.

(a) Drop and Riprap Rocks | (b) Drop and Riprap Raffle

Figure 7.2. Drop Protected with Riprap Rocks.

Manning's roughness, *N*, is directly correlated to the intermediate riprap rock size, D_{50}, in feet as:

$$N = 0.0395 D_{50}^{\frac{1}{6}} \tag{7.1}$$

As recommended (USWDCM 2001), the resistance indicator, *K*, for a stable riprap channel is defined as:

$$K = \frac{US_0^{\ 0.17}}{(S_s - 1)^{0.66}} \tag{7.2}$$

in which S_s = specific gravity of rock such as 2.5 to 2.7. Equation 7.2 is recommended when the flow Froude Number is less than 1.0 or for subcritical flows only. Table 7.1 provides a basis to select a riprap blanket based on the resistance indicator, *K* described by Eq 7.2.

Table 7.1. Commercial Types of Riprap Rock.

Resistance K Value	Riprap Rock Type	Size of D_{50} in Inches
1.5 to 4.0	*VL*	6.0
4.0 to 5.0	*L*	9.0
5.0 to 5.8	*M*	12.0
5.8 to 7.1	*H*	18.0
7.1 to 8.2	*VH*	24.0

In practice, the stream-power based method is good for mild reaches with $Fr \leq 0.8$, and the ratio of water depth to D_{50} is 3 or higher. It takes an iterative process to size the riprap rock. Assign a value for D_{50} in Eq 7.1 to calculate Manning's N. Use Manning's formula to determine the normal flow condition. Check if the product of US_o in Eq 7.2 satisfies the recommended range for the resistance indicator, K. Repeat this procedure until the proper D_{50} is found.

Example 7.1: Continue ***Example 4.1*** in which the channel carries a flow of 2000 cfs with a bottom width of 15 feet, side slope Z=4, and S_o=0.03. The channel bottom slope is improved with drops to S_o=0.0042. Design the riprap protection immediately upstream of the drop structure.

Solution: Let us begin with the V*L type* rock that has D_{50} = 6.0 inches and specific weight of 2.65.

$D_{50} = 6.0\ inch = 0.5\ ft$

$N = 0.0395\ D_{50}^{\ (1/6)} = 0.0352$

Next, the normal flow condition is determined by Manning's formula as: $Y = 6.80\ ft$, $U = 6.96\ fps$ and $F_r = 0.6\ (<0.8)$. Substituting the normal depth and velocity into the rock resistance indicator equation yields:

$$K = \frac{US_0^{\ 0.17}}{(S_s - 1)^{0.66}} = \frac{6.96 \times 0.0042^{0.17}}{(2.65 - 1)^{0.66}} = 1.97 \quad \text{(Ok!)}$$

Figure 7.3 is the recommended thickness of riprap blanket to protect the bank surface, bank toe and channel bed. It is important that an adequate thickness of riprap blanket be provided and large corner rocks be anchored with steel bars. Otherwise the subsurface seepage flows, as shown in Figure 7.4, may scour the foundation of the riprap blanket to result in serious failures.

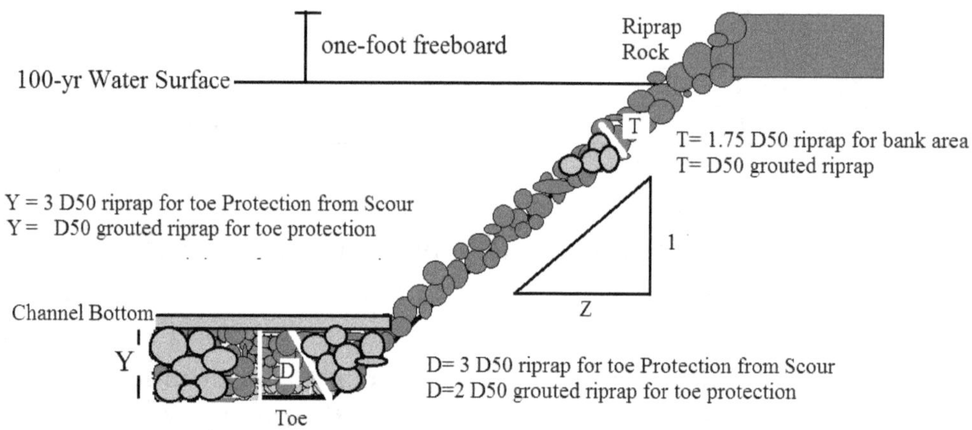

Figure 7.3. Riprap Toe and Bed in Cross Section.

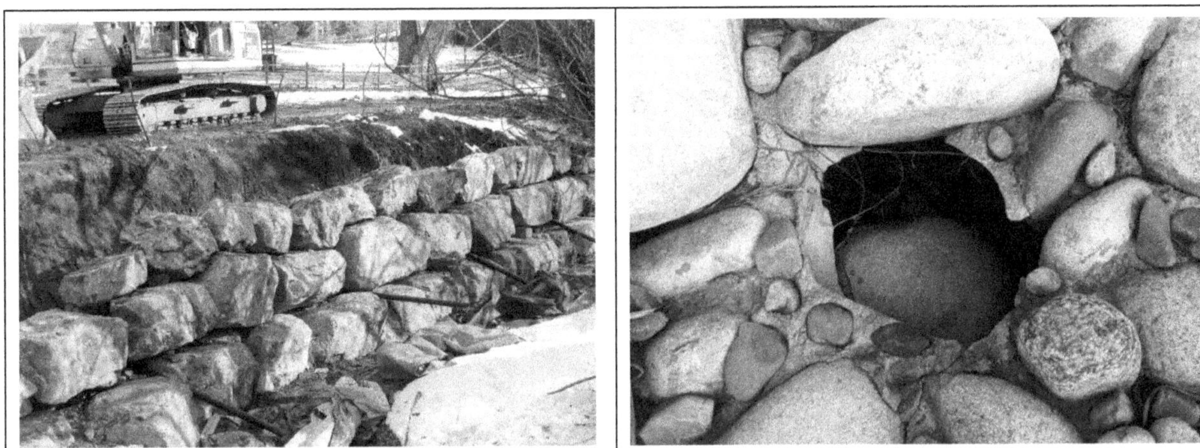

(a) Riprap placed on bank | (b) Failure of rock blanket

Figure 7.4. Steel Anchors and Riprap Blanket.

Example 7.2: The trapezoidal channel in ***Example 7.1*** is laid on a slope of 1.2%. Determine the riprap rock size without a drop structure.

Solution: Set Z=1V:2H and B=15 feet. Try D_{50}=9 inches for Q=2000 cfs as:

DESIGN INFORMATION			
Channel Invert Slope	So =	0.0120	ft/ft
Bottom Width	B =	15.00	ft
Left Side Slope	Z1 =	2.00	ft/ft
Right Side Slope	Z2 =	2.00	ft/ft
Design Flow rate	Q=	2000.00	cfs
Specific Gravity of Rock	Ss =	2.65	
RIPRAP SIZE AND NORMAL FLOW			
Enter Roughness Height	D50=	9.00	inches
Manning N	N=	0.0376	
Normal Flow Depth	**Guess** Y=	6.45	ft
Top Width	T=	40.80	ft
Flow Area	A=	179.95	sq ft
Wetted Perimeter	P=	43.85	ft
Hydraulic Radius	R=	4.10	ft
Flow Velocity	U=	11.11	fps
Hydraulic Depth	D=	4.41	ft
Froude Number	Fr =	0.93	
Difference in Q	**Check** on dQ=	0.00	cfs
Resistance Indicator	K=	3.77	

For this case, the proposed L riprap blanket provides sufficient protection. The flow velocity is 11.11 ft/sec, and the flow Froude number is 0.93<1.0. As shown in Figure 7.5, without a drop, rocks in large sizes are often laid along the channel banks. Over the years, the grass will grow to cover the riprap surfaces.

(a) 24-inch rock for bank protection

(b) Grass seeding on overbank areas

Figure 7.5. Rocks used for Bank Protection.

7.4. SHEAR STRESS-BASED METHOD

7.4.1. Roughness Coefficient

Friction force of water flow in a channel is termed the tractive force that is expressed as the shear stress per unit area. The basis for stable riprap channel design is that the flow-induced shear stress should not exceed the permissible shear stress determined for the lining materials. Based on the logarithmic turbulent flow velocity distribution, Manning's N for riprap lining is directly related to the selected riprap rock size, D_{50}, as (HEC 15):

$$N = \frac{R^{1/6}}{3.82(2.25 + 5.23 Log(R / D_{50}))} \qquad (7.3)$$

Riprap linings are often lined up along the banks for scour protection or paved on the channel bottom to slow down the flow velocity.

(a) Riprap Ling to slow down flow	(b) Riprap Lining to increase depth

Figure 7.6. Composite Riprap Lining.

Dealing with a composite lining surface, as shown in Figure 7.6, Manning's N is weighted by the wetted perimeters as:

$$N = \frac{\sum_{j=1}^{j=m} N_j P_j^{\frac{2}{3}}}{\sum_{j=1}^{j=m} P_j^{\frac{2}{3}}} \tag{7.4}$$

In which N= weighted Manning's N, N_j= *j-th* roughness coefficient, j=*j-th* lining material, m= number of different linings, P=wetted perimeter in [L].

7.4.2. Flow Shear Stress

Using the normal flow as the basis, the flow shear stress applied to the wetted perimeter is equal to:

$$\tau_o = \gamma R S_o \tag{7.5}$$

where τ_0 = flow shear stress in Newton/M^2 or lb/ft^2 on channel bottom and γ = unit weight of water such as 9810 Newton/M^3 or 62.4 lb/ft^3. Along the wetted perimeter in Figure 7.7, the shear stress is reduced from the channel bottom to become vanished on the water surface. The shear stress on the side slopes, τ_S, is linearly proportional to the bottom shear stress, depending on the value of Z as:

$$\tau_S = K_S \gamma R S_o \quad (7.6)$$

The empirical factor, K_S, is defined as:

$$K_S=0.77 \quad \text{for} \quad Z \leq 1.5 \quad (7.7)$$

$$K_S =0.066Z + 0.67 \quad \text{for} \quad 1.5 < Z < 5 \quad (7.8)$$

$$K_S =1.0 \quad \text{for} \quad Z> 5. \quad (7.9)$$

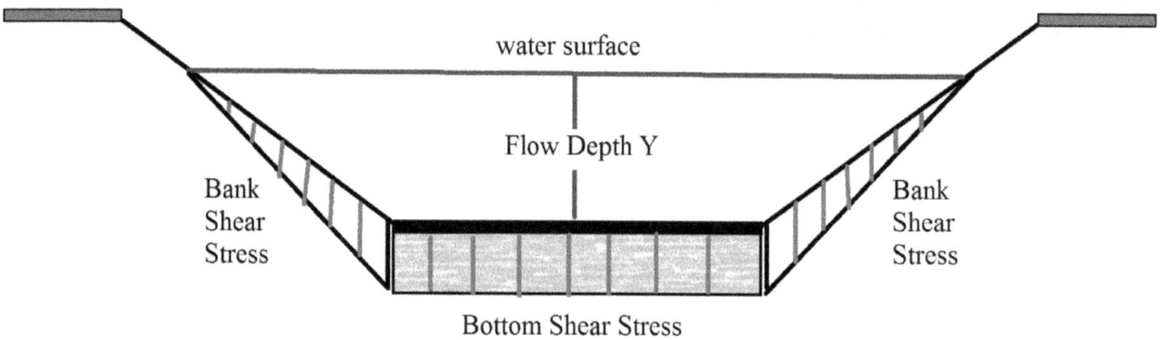

Figure 7.7. Distribution of Shear Stress along Wetted Parameter.

Along a curve reach in Figure 7.8, the outer bank is venerable to bank scours because of the superelevation and increased shear stress. The shear stress on the outer banks, τ_b, is linearly proportional to the bottom shear stress as:

$$\tau_b = K_b \gamma R S_o \quad (7.10)$$

The empirical factor, K_b, is defined as:

$$K_b=2 \quad for \quad (R_c/T) \leq 2 \quad (7.11)$$

$$K_b =2.38-0.206(R_c/T)+0.0073 (R_c/T)^2 \quad for \quad 2 \leq R_c/T \leq 10 \quad (7.12)$$

$$K_b = 1.05 \quad for \quad (R_c/T) \geq 10. \tag{7.13}$$

Where R_c= radius of curvature in [L], and T= top width in channel in [L].

(a) Scour along outer bank	(b) Gabion protection along outer bank

Figure 7.8. Outer Bank Protection Along Curve Reach.

Example 7.3: A trapezoidal channel is built on a slope of 0.015. Its bottom width is B=15 feet and side slope is Z=2. Both side banks are covered with concrete panels while the channel bottom is protected the VL riprap blanket or D_{50}= 6 inches. The water depth is measured to be 5 feet. Determine the discharge and shear stress.

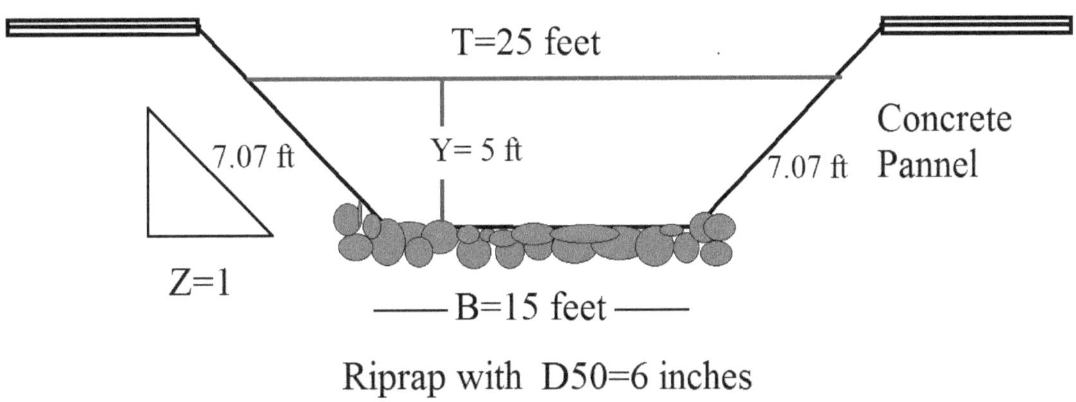

Figure 7.9. Composite Linings.

Solution: For this case, the flow area, wetted perimeter, and hydraulic radius are calculated as:

$$A = \frac{(25+15)}{2} \times 5 = 100 \quad \text{sq feet}$$

$$P = 7.07 + 7.07 + 15 = 29.14 \quad \text{feet}$$

$$R = \frac{A}{P} = 3.43 \quad \text{feet}$$

Using Eq 7.3, Manning's N for D50=6 inches is determined as:

$$N = \frac{3.43^{1/6}}{3.82(2.25 + 5.23Log(3.43/0.5))} = 0.0485 \quad \text{for riprap surface}$$

Manning's N for concrete surface is:

N= 0.014 for concrete surface

Aided with Eq 7.4, the composite Manning's N is calculated as:

$$N = \frac{\sum_{j=1}^{j=m} N_j P_j^{\frac{2}{3}}}{\sum_{j=1}^{j=m} P_j^{\frac{2}{3}}} = \frac{(0.014 \times 7.07^{\frac{2}{3}} + 0.0485 \times 15^{\frac{2}{3}} + 0.014 \times 7.07^{\frac{2}{3}})}{(7.07^{\frac{2}{3}} + 15^{\frac{2}{3}} + 7.07^{\frac{2}{3}})} = 0.0296$$

The discharge for this case is calculated as:

$$Q = \frac{1.486}{N} R^{\frac{2}{3}} \sqrt{S_o} A = \frac{1.486}{0.0296} \times 3.43^{\frac{2}{3}} \sqrt{0.015} \times 100 = 1405 \quad cfs$$

The shear stress for this case is:

$$\tau_o = \gamma R S_o = 62.4 \times 3.43 \times 0.015 = 3.21 \quad \text{pounds/sq ft}$$

The friction force per unit length of channel is:

$F_o = \tau_o P = 3.21 \times 29.14 = 93.5$ pounds per linear foot of reach.

7.4.3. Permissible Shear Stress

The permissible shear stress is set to be the critical shear stress that will induce the incipient motion to individual riprap rock. The permissible shear stress is determined as:

$$\tau_p = F(Ss - 1)\gamma D_{50} \tag{7.14}$$

Where τ_p = permissible shear stress for the channel lining in N/M^2 (lb/ft^2). The value of F is determined by a linear interpolation between the two limits defined by the flow Reynolds number as:

$$F = 0.15 \; if \; Re \leq 2x10^5 \text{ (high limit)} \tag{7.15}$$

$$F = 0.047 \; if \; Re \geq 4x10^4 \text{ (low limit)} \tag{7.16}$$

$$R_e = \frac{UR}{\nu} \tag{7.17}$$

Where R_e = flow Reynolds number, and ν = kinematic viscosity of water. The basic design procedure is to make sure the computed flow shear stress is not exceeding the permissible shear stress determined for the selected riprap lining. To be conservative, a safety factor is introduced to the comparison as:

$$S_f = \frac{\tau_p}{\tau_o} > 1.0 \; to \; 1.5 \tag{7.18}$$

The safety factor, S_f, provides a measure of uncertainty. A safety factor of 1.0 is acceptable for most cases. However, a higher safety factors up to 1.5 is recommended if there exists significant uncertainty regarding the design discharge and/or the consequences of riprap failure lead to huge damage.

Example 7.4: Continue ***Example 4.1*** in which the channel carries a flow of 2000 cfs with a bottom width of 15 feet, side slope Z=2, and S_o=0.03. Design the riprap lining without a grade control.

Solution: Try Type L Riprap Rock with D_{50}=9 inches and S_s=2.65. Solution is presented below:

DESIGN INFORMATION			
Channel Invert Slope	So =	0.0300	ft/ft
Bottom Width	B =	15.00	ft
Left Side Slope	Z1 =	2.00	ft/ft
Right Side Slope	Z2 =	2.00	ft/ft
Design Flow rate	Q=	2000.00	cfs
Specific Gravity of Rock	Ss =	2.65	
RIPRAP SIZE AND NORMAL FLOW			
Enter Roughness Height	D50=	9.00	inches
Manning N	N=	0.0376	
Normal Flow Depth	**Guess** Y=	5.11	ft
Top Width	T=	35.44	ft
Flow Area	A=	128.90	sq ft
Wetted Perimeter	P=	37.86	ft
Hydraulic Radius	R=	3.41	ft
Flow Velocity	U=	15.52	fps
Hydraulic Depth	D=	3.64	ft
Froude Number	Fr =	1.43	
Difference in Q	**Check** on dQ=	0.00	cfs
Resistance Indicator	K=	6.14	
CHECK ON SHEAR STRESS AND K			
Flow Shear Stress	τ_o =	6.37	lb/sq ft
Shear Stress Ratio: Ks	Ks=	0.82	
Flow Side Shear Stress		5.26	lb/sq ft
Flow Reynolds Number	R_e=	4.40E+06	
Value of Variable F	F=	0.15	
Permissible Shear Stress	τ_n =	11.58	lb/sq ft
Shear Safety Factor	S_f=	**1.82**	
Required Safety Factor	S_f-req=	**1.50**	
Riprap or Grouted		**RIPRAP**	

For this case, *Fr=1.43*. The stream power-based method is not suitable for this case because it is a supercritical flow. Therefore, the shear stress-based method is recommended. Type L Riprap Blanket is sufficient to provide a safety factor of 1.82>1.50. In comparison, Example 7.1 presents a VL–riprap blanket with a drop while Example 7.4 presents a L-riprap blanket without a drop. A good engineering practice is to compare the cost and other pros and cons between these alternatives for the final design.

7.5. CLOSING

Sediment transport in a waterway is inevitable. A stable waterway will reach its dynamic balance between sediment erosion and settlement along the stream bed. This Chapter presents the concept of permissible flow velocity and Froude number as a basis to establish the sediment dynamic balance using drop structures. As indicated in Figure 7.10, a drop structure needs riprap protections that can be handed-placed or grouted. A drop structure produces a backwater effect to reduce the energy slope. As indicated in Figure 7.10, between two adjacent drop structures, the floodplains shall be seeded for the grass linings. It is important to select the dry season for construction, and let the grass to be rooted over a winter. It takes a couple of years for grass linings to become stabilized and matured.

(a) Stage of riprap lining

(b) Stage of riprap grouting

(c) Stage of grass seeding

(d) Stage of matured grass and riprap

Figure 7.10. Grass and Riprap Linings.

7.6. HOMEWORK

Q7.1: An existing trapezoidal channel, B=10 feet, Z=4, is built on a slope of 3.0%. This channel will be restored with a stable cross section lined by riprap blankets. The criteria for stable channel is defined by the permissible velocity<= 14.0 fps. (1) Apply the stream power-based method to select a riprap blanket for a design flow of 1200.0 cfs. (2) Repeat the same problem using the permissible shear stress method.

CHAPTER 8
FLOOD CHANNEL DESIGN

In urban areas, floodplains along the major waterways provide open space for parks, bike paths, picnic areas, and recreation uses. Design of urban flood channel shall focus on multiple land uses to maximize the benefits of floodplains. It is important that an urban channel be shaped with multiple layers, including the main channel section for frequent events (6-month to 1-yr events), the overbank areas for minor events (2- to 5-year events), and the entire floodways for major events (10- to 100-yr events). As a result, the concept of multiple events is the major task for urban channel design. Urban channels as shown in Figure 8.1 are aesthetic and also provide significant detention storage volume during the major storm event.

(a) Sport fields on floodplains	(b) River Park along waterway

Figure 8.1. Multiple Land Uses on Urban Channel Floodplains.

As illustrated in Figure 8.2, the main channel of a composite cross section is usually shallow, narrow, hard-bottom with a side slope ranging from 1V:1H to 2H. The capacity of a main channel is sized to carry a low flow up to the 2-year storm event. The overbank areas are usually wide, flat and grass lined to serve as open space. In order to enhance recreational uses, the overbank areas are laid out on a very mild side slope such as 1V:7H to 10H.

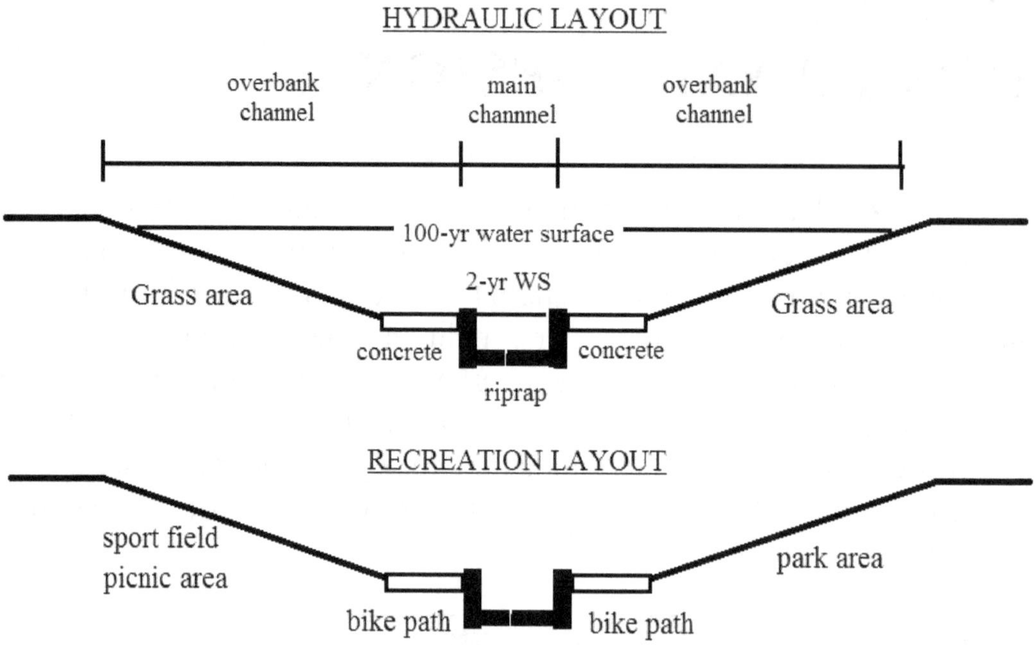

Figure 8.2. Cross Section of Urban Channel.

As shown in Figure 8.3, the floodplain of Cherry Creek in the City of Denver, Colorado provides bike paths and recreational uses during dry weather days. The flow in the main channel is designed to carry up to 300 cfs regulated by the upstream Cherry Creek Dam. During flood events, the entire floodplain, bank-to-bank, is full with storm water. The operation of Cherry Creek is a typical example for urban channels.

(a) Cherry Creek, Denver, Colorado | (b) Flood Flow in Cherry Creek

Figure 8.3. Cherry Creek in Dry Time and Wet Time, Denver, Colorado.

There are three major types of urban channels, depending on the capacity and functions of the main channel. They are: (A) *low flow channel, (B) trickle flow channel,* and *(C) wetland flow channel.*

8.1. LOW FLOW CHANNEL

The boundaries of a *floodplain* in the US are defined by the 100-yr peak flow. To maximize the economic value, a floodplain can be filled with levees and banks from the both sides. The *floodway* is defined by the levees and banks when the water surface is raised by one foot above the 100-yr water surface elevation.

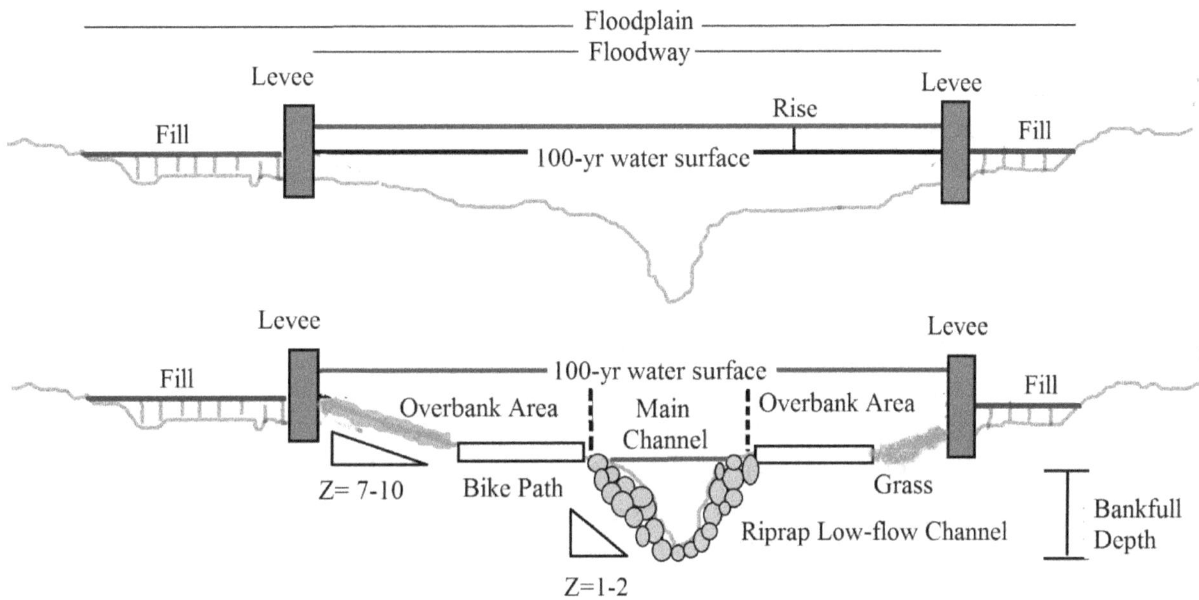

Figure 8.4. Low Flow Channel within Floodway.

A low flow channel as illustrated in Figure 8.4 often runs through the major waterway in an urban area. When the overbank areas are used for recreational trails or bike path as shown in Figure 8.5b, the main channel should have a capacity of passing up to the 2-year peak discharge without a freeboard. The bankfull depth of the main channel ranges from three to five feet. Both banks of the main channel shall be stabilized with riprap or concrete lining protection. Above the main channel are overflow bank areas that are lined with grass on a mild side slope.

(a) Low-flow Concrete Channel | (b) Low-flow Riprap Channel

Figure 8.5. Examples of Low Flow Channel.

8.2. TRICKLE CHANNEL

A trickle channel as shown in Figure 8.6(a) is suitable for river parks and recreational areas on floodplains where return flows from irrigation and base flows from groundwater cause nuisance. A trickle channel is usually laid along the center of the floodplain with a shallow and narrow cross section. The capacity of a trickle channel is about 1.0 to 3.0 percent of the major flow. A composite channel through a river park may consist of multiple layers, including overbank areas for major event, main channel for minor event, and trickle channel for base flows. As expected, a trickle channel is susceptible to bed erosion because it carries nuisance flows continually. It is advisable that the trickle channel be protected by concrete linings or riprap linings with vegetation growth as shown in Figure 8.6(b).

(a) Concrete Trickle Channel | (b) Grass Trickle Channel

Figure 8.6. Examples of Trickle Channel.

8.3. WETLAND CHANNEL

Wetlands are created by the active interaction between surface storm runoff and groundwater. Natural wetlands are located at the low point on a widen floodplain where the water flow is spread into a vegetative area. A wetland area is expected to have a constant water depth 4 to 6 inches. Wherever the existing wetland areas are affected by surface drainage, a wetland channel as shown in Figure 8.7 must be considered. A wetland channel has a similar layout as the low flow channel, except that they provide habitats for aquatic, terrestrial and avian wildlife. Channel erosion protections must not affect the interactions between groundwater and surface water. Wetland channel provides water quality enhancement as the flows move through the marshy vegetation. The abundant bottom vegetation traps sediments and reduces flow velocity. Therefore, it is suggested that a wetland channel be widened to diffuse the flood water and deepened by one or two feet for sediment deposit.

(a) Constructed Wetland Channel	(b) Natural Wetland Channel

Figure 8.7. Examples of Wetland Channel.

The hydraulic roughness of man-made grass lined channels depends on the length and type of grass, as well as the water depth of flow. Table 8.1 is recommended for the determination of hydraulic roughness in composite grass-lined waterways and channels. It is noted that a flow depth exceeding three feet will begin to lay the grass down to form a smoother bottom. Therefore, tall grass does not significantly increase the roughness coefficient to a major runoff flow.

Table 8.1. Manning's Roughness for Composite Grass Channel.

Grass Type	Grass Length	0.7 ft <Depth<1.5 ft	Depth>3.0 ft
Bermuda	2-inch	0.0350	0.0300
	4-inch	0.0400	0.0300
Kentucky	2-inch	0.0350	0.0300
	4-inch	0.0400	0.0300
Any Grass			
Good Stand	12-inch	0.0700	0.0350
	24-inch	0.1000	0.0350
Fair Stand	12-inch	0.0600	0.0350
	24-inch	0.0700	0.0350

8.4. DESIGN CRITERIA FOR COMPOSITE CHANNEL

The following design criteria are recommended for composite channel design:

(A) Channel capacity
A trickle channel is sized to carry the base flow or up to 3% of the peak discharge of the major event. The capacity of a main channel shall be designed not to exceed the two-year event. Often 1/3 to 1/2 of the two year peak discharge is recommended. A wetland channel shall be capable of delivering the existing flow to sustain the habitats in the wide and deep section.

(b) Flow depth
The bankfull depth for a main channel ranges from 1 to 3 feet. The maximum depth in the over bank areas shall not exceed five to seven feet. A wetland channel shall add one to two feet to the flow depth in the widened channel section as the storage volume for future sediment deposit.

(C)Flow velocity
The normal flow velocity for the major design discharge should not exceed 7.0 fps for erosion resistant (clay) soil and 5.0 fps for easily eroded (loamy and dandy) soil. Froude number for the major event in a composite grass channel should not exceed 0.8.

(D) *Freeboard*
No freeboard is needed for the main channel. A freeboard of one foot is recommended for the overbank areas if the flow is a subcritical flow. Superelevation and cross waves in the supercritical flow often require more height of freeboard.

8.5. CONVEYANCE CAPACITY IN COMPOSITE CHANNEL

Due to different types of vegetation on overbank areas, a composite channel is divided into three flow areas: *main channel, left bank area, and right bank area.* The main channel is further divided into the *low flow section* which has a depth less than the bank-full depth, Y_m, and the *high flow section* which submerges the overbank areas. For a given depth, the flow capacity in a composite channel is analyzed under the assumption that the internal friction on the inter-surface is negligible.

8.5.1. Flow in Main Channel

When the flow depth in Figure 8.8 is shallower than Y_m, the flow is confined within the main channel only. The capacity of the main channel at the depth Y is calculated as:

$$A_m = \left[B + 0.5(Z_1 + Z_2)Y\right] \quad (8.1)$$

$$P_m = B + \left[\sqrt{1+Z_1^2} + \sqrt{1+Z_2^2}\right]Y \quad (8.2)$$

$$Q_m = \frac{1.49}{N_m} A_m^{\frac{5}{3}} P_m^{-\frac{2}{3}} \sqrt{S_0} \quad (8.3)$$

$$T_m = B + (Z_1 + Z_2)Y_m \quad (8.4)$$

$$F_{rm} = \left(\frac{Q_m^2 T_m}{g A_m^3}\right)^{0.5} \quad (8.5)$$

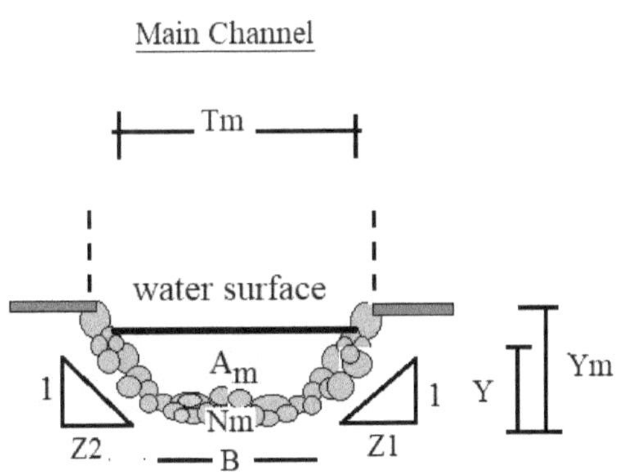

Figure 8.8. Low Flow Channel Capacity.

8.5.2. Flow in Overbank Areas

When the flow exceeds the bankfull capacity of the main channel, water is spilt into the overbank areas. Assuming that the internal frictions on the inter-surfaces are negligible, the capacity of a composite channel with a depth, $Y > Y_m$ as illustrated in Figure 8.9, is separately estimated by the following
s:

(A) Flow in Main Channel

$$A_m = [B + (Z_1 + Z_2)Y_m](Y - Y_m) + [B + 0.5 \times (Z_1 + Z_2)Y_m]Y_m \tag{8.6}$$

$$P_m = B + \left(\sqrt{1 + Z_1^2} + \sqrt{1 + Z_2^2}\right)Y_m \tag{8.7}$$

$$Q_m = \frac{k}{N_m} A_m^{\frac{5}{3}} P_m^{-\frac{2}{3}} \sqrt{S_0} \tag{8.8}$$

$$T_m = B + (Z_1 + Z_2)Y_m \tag{8.9}$$

$$F_{rm} = \left(\frac{Q_m^2 T_m}{g A_m^3}\right)^{0.5} \tag{8.10}$$

(B) Flow in Left Overbank Area

$$A_L = B_L(Y - Y_m) + 0.5 Z_L (Y - Y_m)^2 \tag{8.11}$$

$$P_L = B_L + \sqrt{1 + Z_L^2}(Y - Y_m) \tag{8.12}$$

$$Q_m = \frac{k}{N_m} A_m^{\frac{5}{3}} P_m^{-\frac{2}{3}} \sqrt{S_0} \tag{8.13}$$

$$T_L = B_L + (Y - Y_m)Z_L \tag{8.14}$$

$$F_{rL} = \left(\frac{Q_L^2 T_L}{g A_L^3}\right)^{0.5} \tag{8.15}$$

(C) Flow in Right Overbank Area

$$A_R = B_R(Y - Y_m) + 0.5Z_R(Y - Y_m)^2 \quad (8.16)$$

$$P_R = B_R + \sqrt{1 + Z_R^{\,2}}\,(Y - Y_m) \quad (8.17)$$

$$Q_R = \frac{k}{N_R} A_R^{\frac{5}{3}} P_R^{-\frac{2}{3}} \sqrt{S_0} \quad (8.18)$$

$$T_R = B_R + (Y - Y_m)Z_R \quad (8.19)$$

$$F_{rL} = \left(\frac{Q_L^{\,2} T_L}{g A_L^{\,3}}\right)^{0.5} \quad (8.20)$$

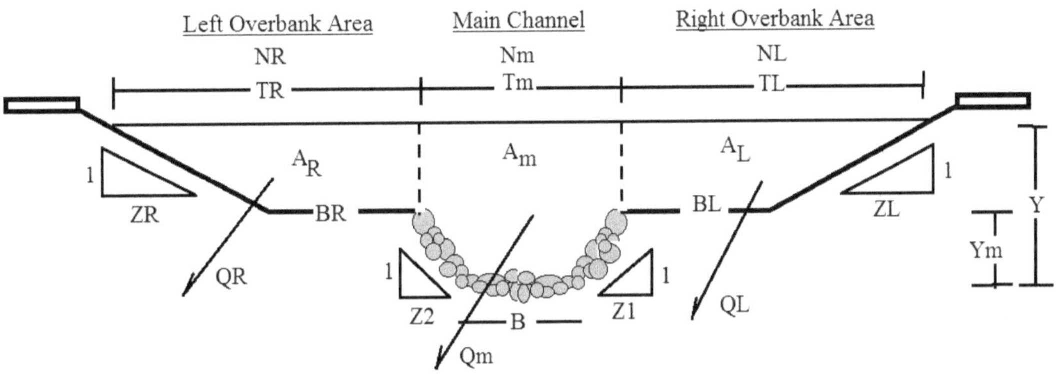

Figure 8.9. Hydraulic Capacity of a Composite Channel.

As a result, the flow parameters for the entire composite channel section are summed as:

$$A = A_L + A_R + A_m \quad (8.21)$$

$$P = P_L + P_R + P_m \quad (8.22)$$

$$Q = Q_L + Q_R + Q_m \quad (8.23)$$

$$T = T_L + T_R + T_m \quad (8.24)$$

$$N = \frac{k}{Q} A^{\frac{5}{3}} P^{-\frac{2}{3}} \sqrt{S_0} \quad (8.25)$$

$$F_r = \left(\frac{Q^2 T}{g A^3} \right)^{0.5} \quad (8.26)$$

in which A = flow area in [L^2], B = bottom width in [L], N = Manning's roughness, T = top width in [L], P = wetted perimeter in [L], S_0 = channel slope in [L/L], Z = channel side slope in [L/L], Q = discharge in [L^3/T], Y = flow depth in [L], and Fr = Froude number. Subscript m represents variables related to the main channel, subscript L represents variables in the left overbank flow area, and subscript R is for the variables in the right overbank flow area.

The rating curve of a channel section describes the relationship between stage and conveyance capacity. Using Eq's 8.1 through 8.26, a rating curve for a composite channel can be established.

Example 8.1: A drainage channel in Figure 8.10 is laid on a slope of 2% and designed to pass 100 cfs in the main channel and 550 cfs by the entire section. The main channel is lined by riprap rocks with a side slope of 1V:1H. The overbank areas are lined by 12-inch mixed grass with a side slope of 1V:5H. Determine the bankfull and maximum water depths for this channel.

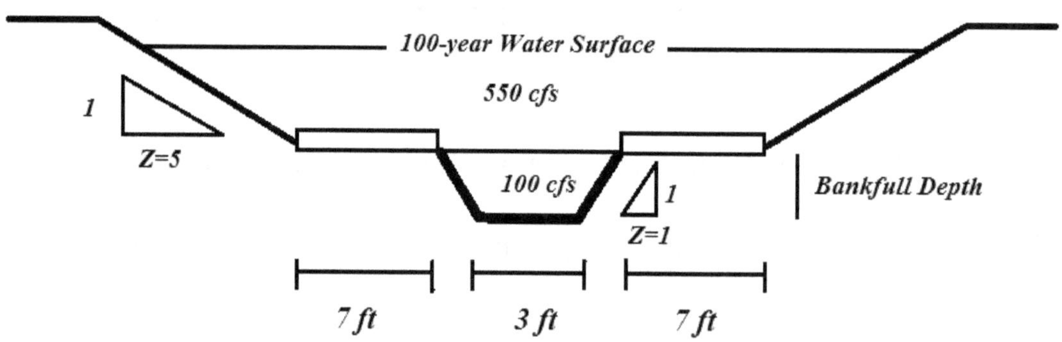

Figure 8.10. Hydraulic Capacity in Composite Channel.

(A) Flow Analysis for Main Channel Section

First, apply the riprap design procedure to confirm that the *VL Type Rock Blanket* with D_{50} of 6.0 inch is adequate for the min channel. The hydraulic roughness in this main channel is 0.0352. The discharge of 100 cfs has a depth of 2.48 ft. For convenience, the main channel for this case is set to have a bankfull depth of 2.50 feet.

(B) Flow Analysis for the Composite Channel Section

According to Table 8.1, the roughness coefficient is found to be 0.07 for 12-inch mixed grass. The capacity of the composite channel at a specified depth can be determined using Eq's 8.6 through 8.20. For example, the following is the calculation of flow rates at a depth of 4.5 feet:

(B.1) Main Channel

The design parameters for the main channel are: $B = 3.0$ *feet*, $N_m = 0.035$, $Z_1 = Z_2 = 1$ *ft/ft*, $Y_m = 2.5$ *feet*, $S_o = 0.02$ *ft/ft*. Aided with Eq's 8.6 through 8.10, the flow parameters are determined as:

$$A_m = [3.0 + (1+1) \times 2.5] \times (4.5 - 2.5) + [3.0 + 0.5 \times (1+1) \times 2.5] \times 2.5 = 29.7 \text{ sq ft}$$

$$P_m = 3.0 + 2.5 \times [(1+1^2)^{0.5} + (1+1^2)^{0.5}] = 10.07 \text{ feet}$$

$$Q_m = \frac{1.49}{0.035} \times 29.75^{\frac{5}{3}} \times 10.07^{-\frac{2}{3}} \times \sqrt{0.02} = 366.65 \text{ cfs}$$

$$T_m = 3.0 + (1+1) \times 2.5 = 8.0 \text{ feet}$$

$$F_{rm} = \sqrt{\frac{366.65^2 \times 8}{32.2 \times 29.75^3}} = 1.126$$

(B.2) Left Overbank Flow

The design parameters for the left overflow section are: $B_L = 7.0$ *ft*, $N_L = 0.07$, $Z_L = 5.0$ *ft/ft*, $S_o = 0.02$ ft/ft. Using Eq's 8.11 through 8.15, we have

$$A_L = 7.0 \times (4.5 - 2.5) + 0.5 \times 5.0 \times (4.5 - 2.5)^2 = 24.0 \text{ sq ft}$$

$$P_L = 7.0 + (4.5 - 2.5) \times \sqrt{1 + 5.0^2} = 17.20 \text{ feet}$$

$$Q_L = \frac{1.49}{0.070} \times 24^{\frac{5}{3}} \times 17.20^{-\frac{2}{3}} \times \sqrt{0.02} = 90.22 \text{ cfs}$$

$$T_L = 7.0 + (4.5 - 2.5) \times 5.0 = 17.0 \text{ feet}$$

$$F_{rL} = \sqrt{\frac{90.22^2 \times 17.0}{32.2 \times 24.0^3}} = 0.558$$

(B.3) Right Overbank Flow

The right overflow section is composed of the following parameters: B_R = *7.0 ft*, N_R = *0.07*, Z_R = *5.0 ft/ft*, S_o = *0.02 ft/ft*. Using Eq's 8.16 through 8.20, we have

$$A_R = 7.0 \times (4.5 - 2.5) + 0.5 \times 5.0 \times (4.5 - 2.5)^2 = 24.0 \text{ sq ft}$$

$$P_R = 7.0 + (4.5 - 2.5) \times \sqrt{1 + 5.0^2} = 17.20 \text{ feet}$$

$$Q_R = \frac{1.49}{0.070} \times 24^{\frac{5}{3}} \times 17.20^{-\frac{2}{3}} \times \sqrt{0.02} = 90.22 \text{ cfs}$$

$$T_R = 7.0 + (4.5 - 2.5) \times 5.0 = 17.0 \text{ feet}$$

$$F_{rR} = \sqrt{\frac{90.22^2 \times 17.0}{32.2 \times 24.0^3}} = 0.558$$

(B.4) **As a result, the flow parameters for the entire section are:**

$$A = A_L + A_R + A_m = 77.75 \text{ sq ft}$$

$$P = P_L + P_R + P_m = 44.47 \text{ feet}$$

$$Q = Q_L + Q_R + Q_m = 549.5 \text{ cfs (close to the design flow of 550 cfs)}$$

$$T = T_L + T_R + T_m = 42.0 \text{ feet}$$

$$N = \frac{1.486}{547.10} \times 77.75^{\frac{5}{3}} \times 44.47^{-\frac{2}{3}} \sqrt{0.02} = 0.043$$

$$F_r = \left(\frac{547.10^2 \times 42}{32.2 \times 77.75^3}\right)^{0.5} = 0.911$$

8.6. CLOSING

Floodplains along an urban channel provide open space for recreational uses, including parks, sports, bike paths, rafting and boating etc. Public safety is always the major concern in the design of any and all hydraulic structures along or across an urban channel. It is critically import that riparian trails are not in a closed system between channel banks. Walkways to get out of the floodplains must be readily available for emergent evacuation. The riparian trail in Figure 8.11 was forced to shut down after loss of human life during an event of flash flood. The entire bike path system in this case needs a re-design to add more exits to the high banks. And signs of flash flood warming must be posted and operated timely to enhance the public safety.

A grade control structure in Figure 8.11 was built across the waterway for multiple purposes. It has a vertical drop for passing the water flow and also a sloping drop to act as a rafting passage. Drop structures must be safely designed to accommodate human's water recreational activities under the low and high flow conditions. Often hydraulic structures were designed to achieve their hydraulic purposes, but they do need habitations to accommodate other needs after years in use.

(a) Vertical and Sloping Composite Drop

(b) Trail between Banks without Exit

Figure 8.11. Public Safety on and along Urban Floodplains.

8.7. HOMEWORK

Q8.1: The drainage channel in Figure HW 8.1 is laid on a slope of 2%. The 100-yr peak discharge is 550 cfs. The main channel has Manning N=0.035 with a side slope of 1V:1H. The overbank areas have Manning N=0.07 with a side slope of 1V:5H. The bankfull depth is set to be 2.5 feet. Determine the encroachment distance, X in Figure HW8.1, if the 100-yr water depth can be raised by 0.5 foot.

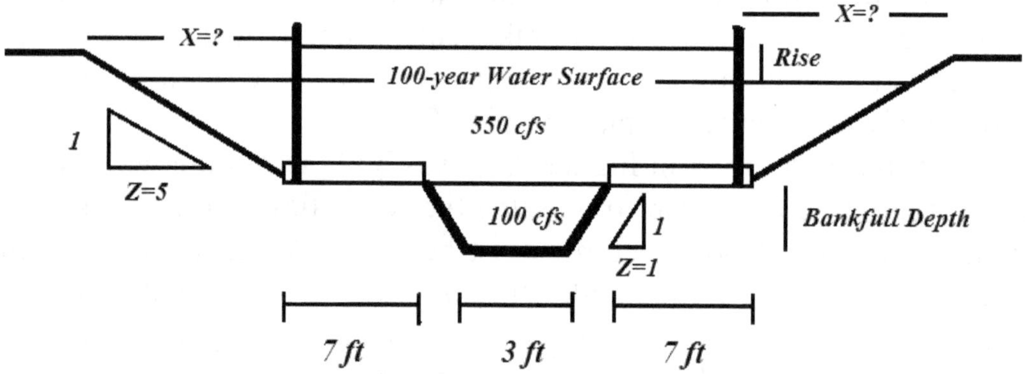

Figure HW8.1. Floodplain Encroachment.

Q8-2: Figure HW8.2 illustrates the top view of a flood channel. This channel starts with a straight reach and then turns into a curved reach with a radius of 250 feet. The outfall reach is straight and drains into a lake. This channel is to be lined by grass with drops. The design discharge for a major storm event is 1200 cfs, and the trickle channel is designed to carry 20 cfs. The design parameters for this channel are:

1) The channel cross section shall be trapezoidal with B=10 feet and Z=4.0,
2) The channel will be lined by Type C grass,
3) The permissible velocity in the channel is 7 fps with its Froude number<0.70,
4) The upstream of a drop needs to be protected by riprap blankets,
5) The downstream of a drop shall be equipped with a plunging pool, and
6) Add a freeboard of 6-inch to take care of super-elevation.

The elevations along the centerline of the channel reach are given as:

Section	Ground Elevation (ft)	Reach	Length (ft)
1	4,995	1	400
2	4,988	2	400
3	4,983	3	400
4	4,980	4	in Lake

Your Task: Design drops at Station 0+400 ft and Station 0+800

a) Propose a new slope using a drop
b) Determine the drop height for design flow Q=1200 cfs
c) Size the top weir to pass a low flow Q-low= 20 cfs
d) Feature the plunging pool downstream of the drop
e) Choose a riprap blanket to protect the approach channel upstream of the drop
f) Design the grass channel using Type C grass
g) Add a freeboard
h) Discuss how to add the maintenance access ramp

Design the riprap channel from Station 0+800 ft and Station 0+1200 using the shear-stress method and discuss the backwater effects from the lake.

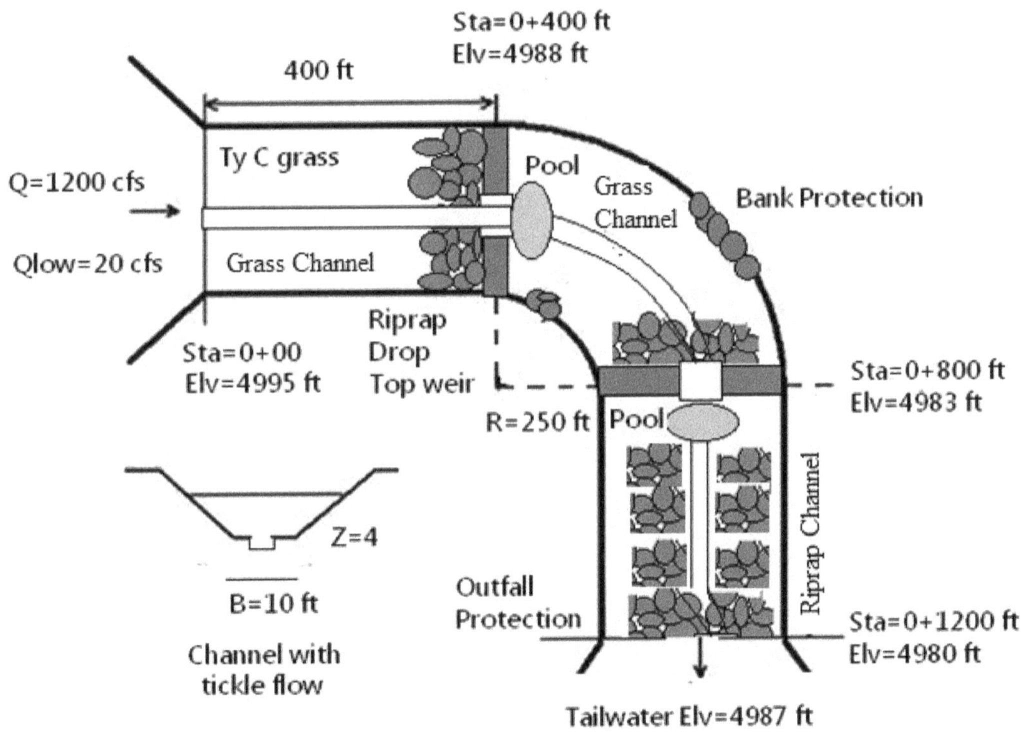

Figure HW8.2. Layout of Flood Channel.

CHAPTER 9
HIGH GRADIENT CHANNEL

The open channel flow on a straight and steep slope tends to be fast and shallow. The friction from the solid walls has a significant influence on the flow velocity. As a result, the flow velocity on the water surface may become less than its wave speed. Consequently, the uniform flow predicted by Manning's equation breaks into a train of traveling waves or pulses as a response to the temporary force balance among gravitational force, skin friction, and internal turbulence. This phenomenon is termed *slug flow* or *pulsating flow*. Roll waves, as shown in Figure 9.1a and b, in a pulsating flow exhibit similar characteristics to a moving jump. They are the result of transition from a wall boundary layer flow to a turbulent state of flow (Chow 1959). After developing a pulsating flow in a steep channel, roll waves will progress downstream and eventually break and form hydraulic bores or shock waves with rough tumbling heads and smooth tails. These roll waves will spill water out of the banks if the channel does not have an additional height of freeboard.

(a) Roll Waves in Chute | (b) Roll Waves in Channel

Figure 9.1. Roll Waves in High Gradient Channel.

It is clear that the uniform flow approach does not adequately address waves and energy dissipation in a supercritical flow. A channel designed by the uniform flow approach on a high gradient tends to be wide and shallow. Such a channel cross section must be equipped with a sufficient freeboard to prevent water from being spilled out at a curve or jumps at a bend.

9.1. STABILITY OF SUPERCRITICAL FLOW

The challenge in the design of a high gradient channel is directly related to the interchange between the flow depth and the velocity head. A supercritical flow is sensitive to channel alignment, cross sectional geometry, surface roughness, and channel invert slope. Often the flow velocity in a concrete channel is as high as 20 to 30 feet/sec or a velocity head of 6 to 14 ft. The major concern in the design of high gradient channel is to make sure that the channel cross section is sufficiently deep to accommodate water surface waves and oblique jumps. As recommended by Escoffier and Boyd in 1962, the flow Froude number defined as the ratio of flow velocity to wave celerity may serve as a basis to divide the open channel flows into subcritical, critical, and supercritical flow regimes. Furthermore, the ratio of incremental flow velocity to incremental wave celerity serves as another basis to divide the high Froude number flows into stable and unstable flow regimes. The instability of a supercritical flow depends on the existence of roll waves which is also termed pulsating flow or slug flow.

Vedernikov number was developed to identify the existence of a pulsating flow in a high gradient channel (Vedernikov 1945 and 1946). Vedernikov number is defined as:

$$N_v = m(1 - R\frac{dP}{dA})F_r \tag{9.1}$$

$$F_r = \sqrt{\frac{TQ^2}{gA^3}} \tag{9.2}$$

in which m = *2/3* when Manning's formula is used or *1/2* when Chezy's formula is used, R = hydraulic radius in [L], P = wetted perimeter I [L], F_r = flow Froude number for the design flow, Q, in [L^3/T], T= top width in [L], g=gravitational acceleration in [L/T^2] and A = flow area in [L^2]. To be a stable uniform flow, N_v shall be less than or equal to unity. To apply Eq 9.1 to a trapezoidal channel section in Figure 9.2, the parameters are derived as:

$$R = \frac{A}{P} = \frac{y(b+zy)}{b+2ky} \tag{9.3}$$

$$k = \sqrt{1+z^2} \tag{9.4}$$

$$\frac{dP}{dA} = \frac{2k}{b+2zy} \tag{9.5}$$

in which y = flow depth in [L], b = channel bottom width in [L], and z = channel side slope in [L/L].

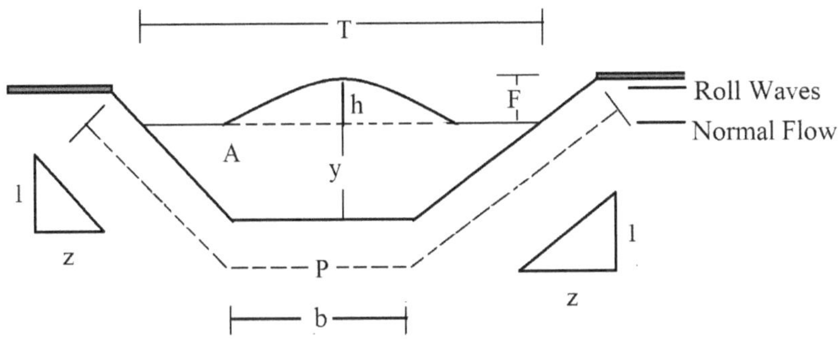

Figure 9.2. Cross Sectional Element in Channel Flow.

Normalize Eq's 9.3 through 9.5 by channel bottom width, b. Next substituting Eq's 9.3 and 9.4 into Eq 9.5 yields the limiting Froude number, F_L, as:

$$F_L = \frac{3}{2}\left[\frac{(1+2kY^*)(1+2zY^*)}{1+2zY^* + 2kzY^{*2}}\right] \tag{9.6}$$

$$Y^* = \frac{y}{b} \tag{9.7}$$

Equation 9.6 is the definition of *limiting Froude number* for having a stable uniform supercritical flow in a trapezoidal channel. In other words, when the flow Froude number is greater than the limiting Froude number, the flow becomes unstable. When $z = 0$, Eq 9.6 is reduced to:

$$F_L = \frac{3}{2}(2Y^* + 1) \tag{9.8}$$

Figure 9.3 presents the family curves for limiting Froude numbers with z = *0.0, 1.0, 2.0 and 3.0*. These curves are converged at $F_L = 1.5$. It implies that supercritical uniform flows can be sustained in trapezoidal channels when the flow Froude number, $F_L < 1.5$, otherwise roll waves will be developed. Applying the limiting Froude number as a criterion, flows with Froude number > *1.5* are further divided into two categories: (1) *stable uniform flows* when flow Froude number: $F_r \leq F_L$, and (2) *pulsating flows* when $F_r > F_L$.

As a rule of thumb, a deep and narrow channel tends to carry stable flows while a shallow and wide channel tends to carry pulsating flows. Care has to be taken when designing a steep channel. The channel cross sectional geometry in terms of *y/b* ratio, channel slope, and lining roughness shall be selected to avoid roll waves or to add an adequate freeboard if the roll waves will be developed. Otherwise an additional freeboard section has to be provided.

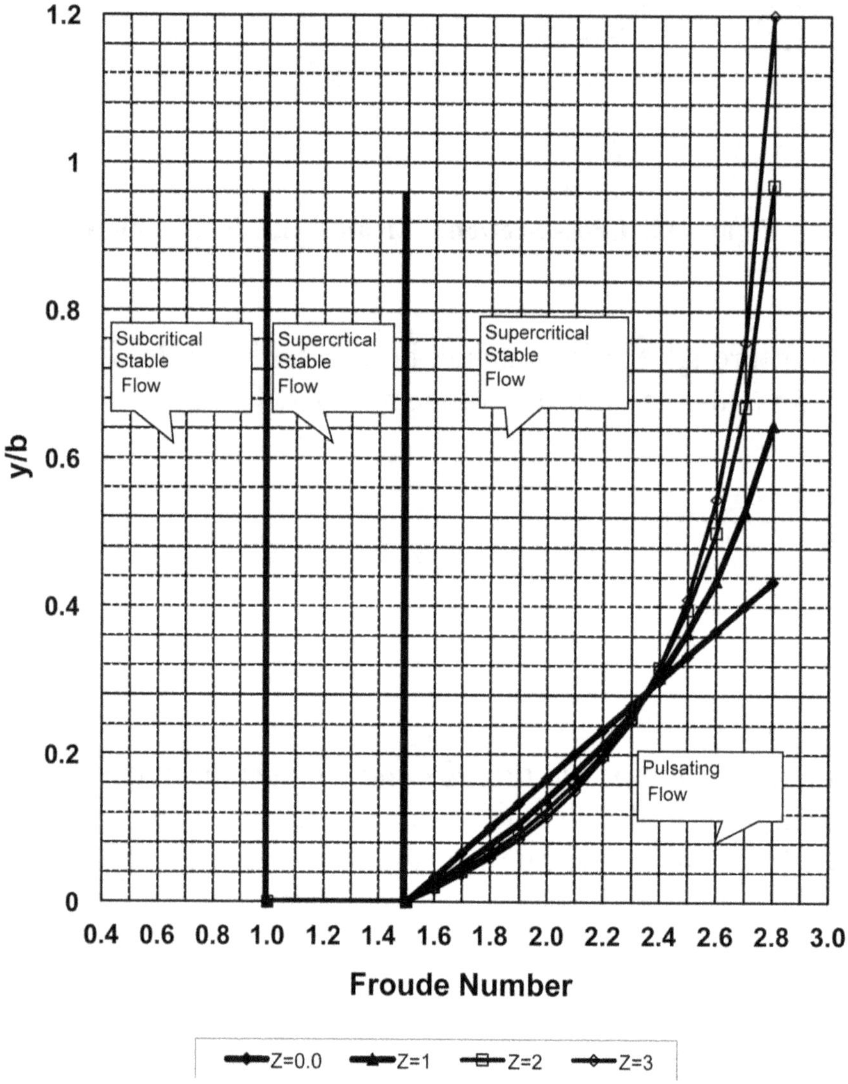

Figure 9.3. Flow Stability Versus Limiting Froude Number.

Example 9.1: A concrete channel is built on a slope: S_o=3%. The channel cross section has a bottom width, b=15 ft, and side slope, z=1. Examine if the design flow of 2500 cfs is stable or unstable in this channel.

Solution:

Given Design Information:				
	Bottom Width	b=	15.00	feet
	Side Slope	z=	1.00	ft/ft
	Manning's Roughness	N=	0.014	
	Longitudinal Slope	S=	0.03	ft/ft
	Design Flow	Q=	2500.00	cfs
Normal Flow Condition				
	Normal Flow Depth	Yn=	3.71	feet
	Wetted Perimeter	P=	25.50	ft/ft
	Normal Flow Area	An=	69.49	sq ft
	Normal Flow Velocity	Vn=	35.97	ft
	Hydraulic Radius	R=	2.72	feet
	Top Width	T=	22.43	feet
	Flow Fourde Number	Fr=	3.60	

Stability of Flow: Stable or Pulsating

Flow Depth y=Yn feet	Channel Width b feet	Side Slope z ft/ft	Parameter k	Ratio Y*=y/b	Limiting Froude Number FL	Flow Froude Number Fr	Channel Flow Condition
3.71	15.0	1.00	1.4142	0.2476	2.29	3.60	**Unstable**

Example 9.2: A rectangular concrete channel is designed to carry 1500 cfs on a slope of 2.0%. Consider Manning's N = 0.014. Determine the maximum b/y ratio to be a stable flow.

Solution:

1. Selection a range of channel widths: b= 5, 7, 10, 12, 15, and 20 feet
2. Calculate the normal flow and flow Froude number using Manning's formula for each width.
3. Calculate the limiting Froude number to compare with the flow Froude number

As shown in Table 9.1, the channel will carry a stable flow at 1500 cfs when the channel with is less than or equal to 13 feet. Wider than 13 feet, the flow becomes unstable.

Table 9.1. Flow Stability Analysis for Q=1500 cfs, N=0.014, So= 2.0%.

Width b 'ft	**Depth 'y ft**	**Flow Froude No. F_r**	**Limiting Froude No. F_L**	**Comment**
13	4.13	2.42	2.48	stable
14	3.87	2.48	2.33	unstable
15	3.65	2.53	2.23	unstable
20	2.90	2.67	1.94	unstable

9.2. ROLL WAVES

Roll waves are formed with a continuous amplification of a small perturbation. For the purpose of design, this complicated process is simplified by the model of positive surges that have an advancing front with the profile similar to a moving hydraulic jump. When the height of the surge is small, the surge appears like an undular jump. When the height is increasing, the undulation will eventually disappear and the surge will have a sharp and steep front. The moving hydraulic jump was investigated by Liggett (1975) to solve the gradually varied flow equation, and achieved the same criteria of instability as Vedernikov number. This conclusion further confirms the dynamic similarity between moving hydraulic jumps and roll waves. In 1999, Guo presented a moving hydraulic jump to model roll waves. The wave front is considered as a positive surge. The unsteady flow pattern can be converted to its steady pattern by a coordinate system moving with the bore velocity. As illustrated in Figure 9.4, the continuity principle is:

$$(V_2 - V_w)A_2 = (V_1 - V_w)A_1 \qquad (9.9)$$

Re-arranging Eq 9.9 yields:

$$V_2 = \frac{(V_1 - V_w)A_1 + V_w A_2}{A_2} \qquad (9.10)$$

And the momentum principle is:

$$(V_w - V_2)(V_2 - V_1) = (\bar{y}_2 - \frac{A_1}{A_2}\bar{y}_1)g \qquad (9.11)$$

in which Vw = wave velocity in [L], V = flow velocity in [L/T], A = flow area in [L^2], and $\bar{y}$ = distance in [L] to the centroid of the flow area, approximated by 0.5y. The subscript 2 denotes the design condition including roll waves, and the subscript 1 represents the limiting condition defined by Vedernikov number.

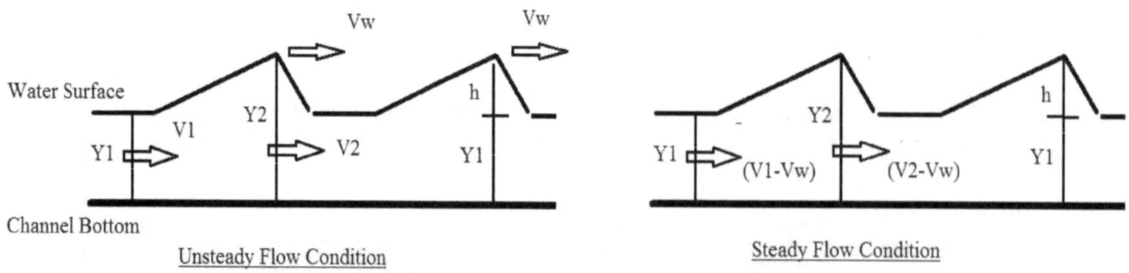

Figure 9.4. Moving Hydraulic Jump.

Solving Eq's 9.9, 9.10, and 9.11 simultaneously yields

$$V_w = V_1 + \sqrt{\frac{(A_2\bar{y}_2 - A_1\bar{y}_1)g}{A_1(1 - A_1 / A_2)}} \tag{9.12}$$

Let the height of roll waves be the difference between flow depths as

$$h = y_2 - y_1 \tag{9.13}$$

$$C = V_w - V_2 \tag{9.14}$$

in which h = height of roll waves in [L], and C = wave celerity in [L/T]. In a high gradient channel, the water surface in a cross section is fluctuating due to cross waves (Stonstreet, 1996). Considering that the roll waves near the center of the channel section is similar to that in a rectangular channel, the representative height of roll waves can be derived by simplifying Eq's 9.10 and 9.11 for rectangular channel. Canceling the wave velocity in Eq's 9.10 and 9.11, the height of roll waves is derived as (Guo, 1999):

Equation 9.15 is only applicable if $F_2 > F_1$. Otherwise h=0.

$$F_1 = \frac{V_1}{C} \tag{9.15}$$

$$F_2 = \frac{V_2}{C} \tag{9.16}$$

In practice, F_2 is the flow Froude number for the design discharge and F_1 is the limiting Froude number determined by Vedernikov's number. When the height of roll wave is small compared with the depth of flow, i.e. $y_1 = y_2$, Eq 9.14 is reduced to:

$$h = \frac{C^2}{g}\left(\frac{V_2}{C} - \frac{V_1}{C}\right) = \frac{(V_w - V_2)^2}{g}(F_2 - F_1) \tag{9.17}$$

Equation 9.17 agrees with the two-dimensional surge model (Chow, 1959) and can provide an estimated wave height of the spectrum of roll waves. To apply the above procedure to the design of a high gradient channel, the design condition, at first, shall be evaluated by Eq 9.6 or Figure 9.3. If the design condition exceeds the limiting condition, a pulsating flow is expected. The height of roll waves can be estimated by Eq 9.14 for the freeboard design.

Example 9.3: Continued with Example 9.1. With Q=2500 cfs, b=15 feet, z=1.0, and S_o=3%. Determine the height of roll waves.

Solution: Let Section 1 carry the stable normal flow condition as:

$$y_1 = Y_n = 3.71\ ft,\ \ V_1 = V_n = 35.97\ ft/s$$

$$A_1 = \frac{(B+T_1)}{2} y_1 = \frac{15+(15+2\times 3.71)}{2}\times 3.71 = 69.49\ ft^2$$

Because the normal flow condition is stable, its flow Froude number is set to be: $F_1 = F_L = 2.29$. Next assume that *h=2.96 ft.* The flow condition at Section 2 includes:

$$y_2 = y_1 + h = 3.71 + 2.96 = 6.67\ ft$$

$$A_2 = \frac{(B+T_2)}{2} y_2 = \frac{15+(15+2\times 6.67)}{2}\times 6.67 = 144.03\ ft^2$$

$$\overline{y_2} \approx \frac{y_2}{2} = 3.34\ ft,\ \ \overline{y_1} \approx \frac{y_1}{2} = \frac{Y_n}{2} = 1.86\ ft,$$

The wave velocity is calculated as:

$$V_w = V_1 + \sqrt{\frac{(A_2\overline{y_2} - A_1\overline{y_1})g}{A_1(1 - A_1/A_2)}} = 53.73\ fps$$

$$V_2 = \frac{(V_1 - V_w)A_1 + V_w A_2}{A_2} = 45.2\ fps$$

At Section 2, the flow is unstable, therefore the flow Froude number is set to be: $F_2 = F_r = 3.60$.

$$C = V_w - V_2 = 53.37 - 45.2 = 8.53\ fps$$

The computed h=2.13 ft is different from the assumed, h=2.96 ft. We need to refine the solution till the error becomes negligible. The solution for this case is:

GUESS		Flow Condition Before BORE				
Wave height h		Flow Depth $y_1 = Y_n$	Flow Area A_1	Centroid Depth $\bar{y}_1$	Flow Velocity V_1	Limiting Froude Number F_L
feet		feet	ft^2	feet	ft/s	
2.39		3.71	69.49	1.86	35.97	2.29

Flow Condition During BORE						Wave Celerity	CHECK Computed Wave Height
Flow Depth y_2	Flow Area A_2	Centroid Depth $\bar{y}_2$	Wave Speed V_w	Flow Velocity V_2	Flow Froude Number F_r	C	
feet	ft^2	feet	ft/s	ft/s		ft/s	feet
6.10	128.84	3.05	52.28	43.48	3.60	8.79	2.39

9.3. CLOSING

High gradient channels are commonly acceptable for the sites where the right of way is limited. On a steep slope, the channel is operated at a high hydraulic capacity to deliver flows. The design procedure for a steep channel is to refine the normal flow depth with a sufficient freeboard. For a steep channel, the excavated depth is equal to:

$$D = Y_n + F \tag{9.18}$$

in which D = channel depth in [L], F= height of freeboard in [L], and Y_n = normal flow depth in [L] determined by Manning's equation.

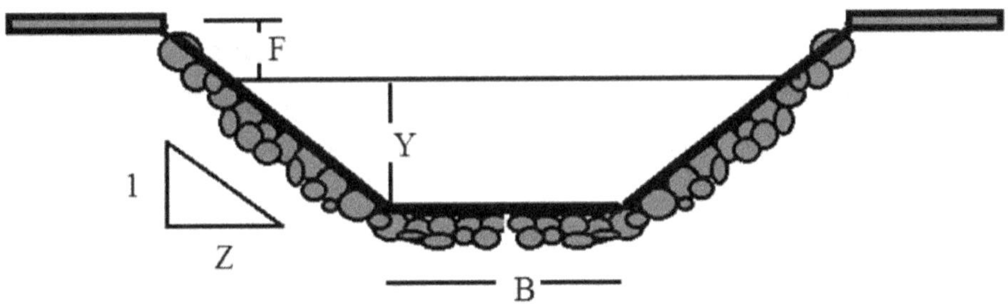

Figure. 9.5. Channel Excavated Area.

For a supercritical flow, the height of freeboard is recommended to be:

$$F = \max[\frac{1}{6}(Y_n + \frac{U_n^2}{2g}), F_H, h] \text{ for supercritical flow in a straight reach} \quad (9.19)$$

in which F_H= 3.0 feet, U_n= normal flow velocity in [L/T] for the design discharge, and g= gravitational acceleration in [L/T^2]. Equation 9.19 presents a conservative approach to warrant a sufficient channel cross section for high Froude number flows in the channel.

9.4. HOMEWORK

Q 9.1: A mining channel is built to carry a flow of 2500 cfs on a slope of 3% with Manning's N=0.025 (rough concrete surface) and side slope z=1v:1H. Since the site is wide open, the channel width may vary from b=15 to 25 feet. Evaluate the flow condition for channel width: b=15 or 25 feet.

Solution:

Width	Depth	Limiting	Flow	Comment
b	y	Froude No	Froude No	
feet	feet	FL	Fr	
15.00	5.18	2.47	2.08	Stable
25.00	3.91	2.06	2.10	Unstable

Q 9.2: Continue with Q9.1. Determine the heights of roll waves and freeboard for the case with b=25 feet (Answer: h=0.16 ft). If the channel construction cost is linearly proportional to the excavated cross sectional area, recommend a least cost cross section for this case.

CHAPTER 10
CULVERT HYDRAULICS

A culvert is often laid under a highway in order to maintain the continuity of water flows. The hydraulic capacity of a culvert varies with respect to the water depth at the entrance. As shown in Figure 10.1, a culvert structure includes *inlet*, *pipe barrels*, and *outlet*.

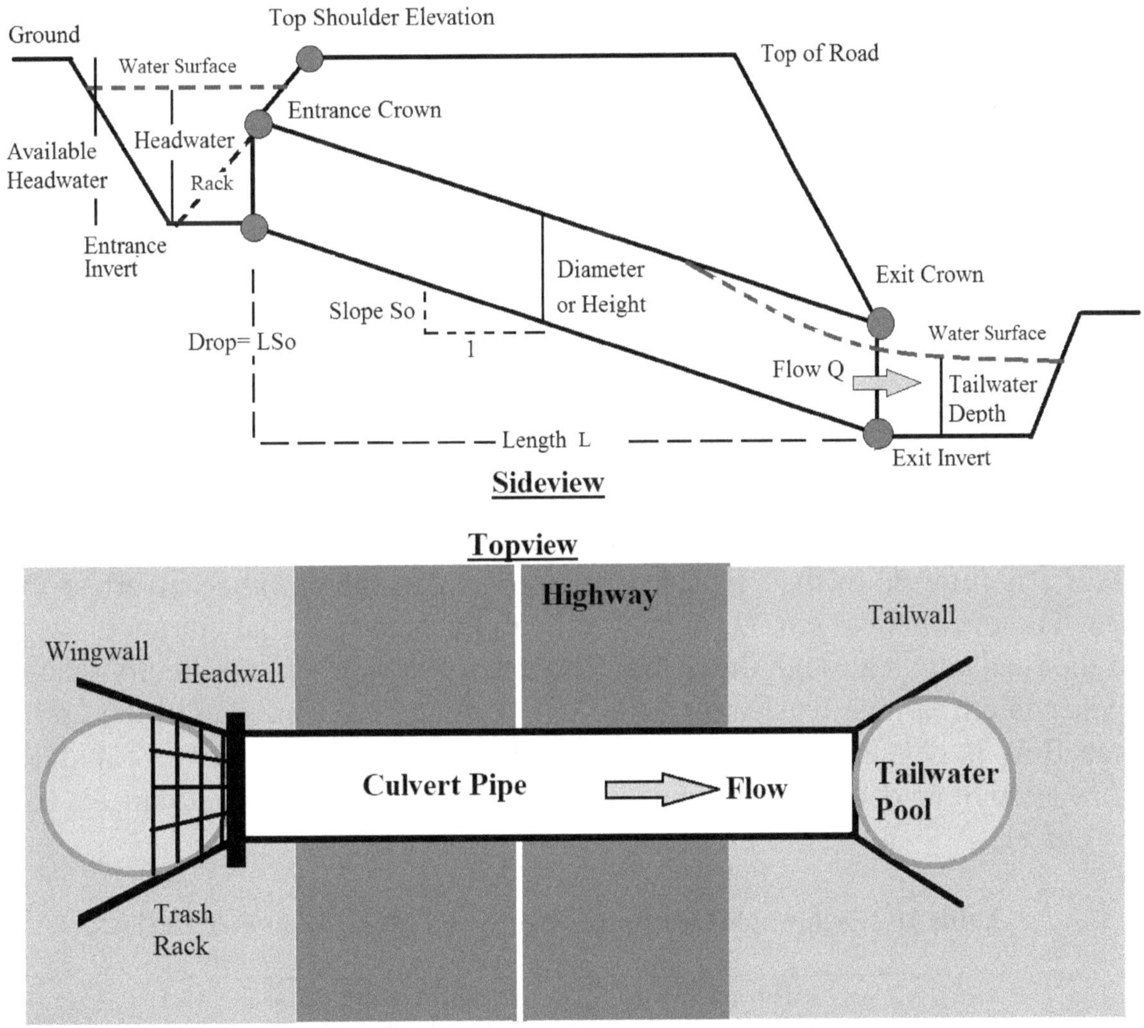

Figure 10.1. Elements of Culvert.

Design of culvert inlet takes into consideration of hydraulic efficiency, structural safety, and long-term maintenance, including urban trash, debris, sediment deposit. For instance, head wall is built around the culvert entrance, wing walls are built to guide the water flow to enter the culvert entrance, and tail walls are placed at the culvert exit to safely spread out the concentrated flow into the downstream floodplains.

(a) On-stream Culvert

(b) Crossing Culvert

Figure 10.2. Examples of Culverts.

10.1. CULVERT DESIGN CONSIDERATIONS

As a culvert is laid along the natural stream that crosses a highway, its alignment shall be located as close to the natural waterway as possible. At the entrance, the *headwater depth,* H_w, is referred to the vertical distance between the entrance invert to the water surface. The *available depth,* H_a, is the vertical distance from the entrance invert to the top of the roadway. Although the hydraulic capacity of a culvert is directly related to the headwater depth, a high headwater depth may damage the pipe due to the uplift as the seepage flow is developed beneath the pipe line. To be conservative, Table 10.1 presents the recommended allowable headwater depth to circular pipe diameter (or height of box culvert).

Table 10.1. Allowable Headwater Depth to Culvert Diameter Ratios.

Culvert Diameter or Height in feet	**Maximum Ratio Headwater/Diameter Hw/d**
less than 3	1.5
from 3 to 5	1.3
from 5 to 7	1.2
larger than 7	1.0

In operation, trash and debris carried in stormwater impose the risk of pipe clogging at the entrance. Table 10.2 presents the recommended minimum culvert sizes for various applications. The length of culvert pipe used for highway crossing shall be in multiples of two feet and the length of a concrete box culvert shall be to the nearest foot.

Table 10.2. Recommended Minimum Culvert Sizes.

Type of Culvert	Minimum Diameter
Cross Culvert	24 inches
Side Drain	18 inches
Median Drain	18 inches
Storm Sewer Trunk Line	18 inches
Storm Sewer Connections	15 inches
Irrigation Crossing	18 inches

A concrete *headwall* in Figure 10.3a is recommended for culverts with 50 square feet of crossing area (>96-inch in diameter) and larger. The headwall should be placed perpendicular to the culvert centerline. All concrete box culverts in Figure 10.3a and pipes with full concrete headwalls require concrete *wingwalls*. Wing walls shall maintain a streamlined angle between 45 to 75-degrees, to guide the water into the entrance of the culvert. Concrete aprons in Figure 10.3b are recommended to be placed at the culvert exit for scour protection.

(a) Head Wall and Wing Walls

(b) Outlet Apron

Figure 10.3. Culvert Elements.

10.2. CULVERT SIZING

Although a culvert is sized to pass the design flow, the service of the culvert is subject to a wide range of flows. Therefore, the concept of multiple frequencies (events) is important to culvert design. The barrel (conduit) is often sized based on the assumption of normal flow for the design flow. The normal flow in a partially full circular pipe is described as:

$$A = \frac{d^2}{4}(\theta - \sin\theta\cos\theta) \tag{10.1}$$

$$P = d\theta \tag{10.2}$$

$$R = \frac{d(\theta - \sin\theta\cos\theta)}{4\theta} \tag{10.3}$$

$$Y = \frac{d}{2}(1 - \cos\theta) \tag{10.4}$$

$$T = d\sin\theta \tag{10.5}$$

$$Q = \frac{k}{N}P^{-\frac{2}{3}}A^{\frac{5}{3}}\sqrt{S_0} \tag{10.6}$$

$$V = \frac{Q}{A} \tag{10.7}$$

$$F_r = \sqrt{\frac{Q^2T}{gA^3}} \tag{10.8}$$

in which A = flow area in [L^2] , d = diameter of the pipe in [L], P = wetted perimeter in [L], R = hydraulic radius in [L], θ = a half of the central angle shown in Figure 10.4, k=1.486 for ft-second units or 1.0 for meter-second units and Y = flow depth in [L]. It is noticed that when the pipe is full, the top width of the flow area is reduced to zero or Froude number for flowing full flow vanishes.

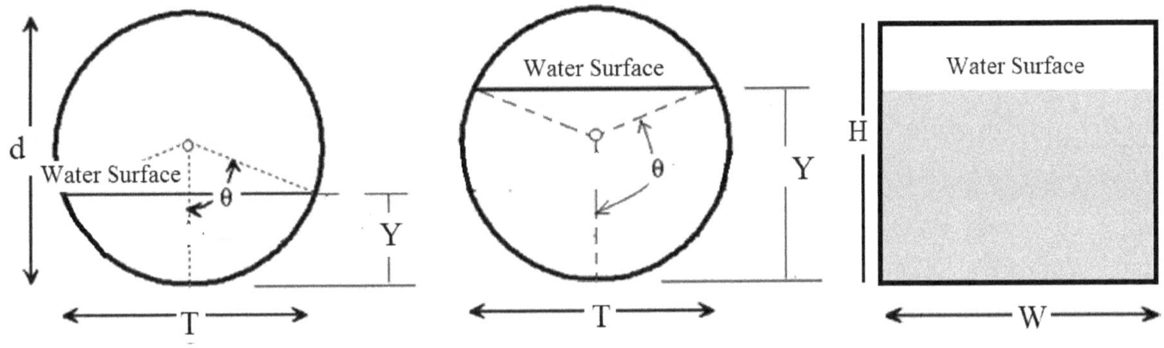

Figure 10.4. Hydraulic Parameters in a Closed Conduit.

For a box conduit in Figure 10.4 with a height of H in [L] and width of W in [L], its hydraulic radius is computed by:

$$A = WY \tag{10.9}$$

$$P = W + 2Y \tag{10.10}$$

$$R = \frac{WY}{W + 2Y} \tag{10.11}$$

When both entrance and exit are submerged, the flow in the culvert is pressurized. The flowing full hydraulic radius of a circular pipe becomes:

$$R = \frac{A}{P} = \frac{d}{4} \tag{10.12}$$

With a half of central angle of 180 degrees, Eq 10.6 is reduced to:

$$d = \left(\frac{NQ}{K\sqrt{S_0}}\right)^{\frac{3}{8}} \tag{10.13}$$

in which N = manning roughness, Q = design flow in $[L^3/T]$, S_o = conduit invert slope, d = hydraulically required circular diameter in [L], and K = 0.462 for feet-second units or 0.311 for meter-second units. Equation 10.13 provides a hydraulically required pipe diameter in feet or meter. For the purpose of design, the next larger commercially available pipe shall be selected, and the normal flow condition for the design flow in the selected pipe is solved by the trial-and-error to find the proper central angle.

Example 10.1: Determine the normal flow condition in a 24-inch concrete pipe laid on a slope of 1%.

Solution:

Pipe Diameter	D =	2.00	ft	0.62	m				
Pipe Slope	S_o =	0.0100	ft/ft	0.0100	m/m				
Manning's Roughness	N =	0.0140		0.0140					

A half Central Angle degree	Angle radian	Depth Y ft	Area A ft^2	Wetted P P ft	H Radius R ft	Top Width T ft	Velocity V fps	Flow Q fps	Froude Fr
0.00	0.00	0.00	0.00	0.00	0.00	0.00	0.0	0	NP
30.00	0.52	0.13	0.09	1.05	0.09	1.00	2.1	0.2	1.22
60.00	1.05	0.50	0.61	2.09	0.29	1.73	4.7	2.9	1.39
104.85	1.83	1.26	2.08	3.66	0.57	1.93	7.3	15.1	1.24
120.00	2.09	1.50	2.53	4.19	0.60	1.73	7.6	19.2	1.11
150.00	2.62	1.87	3.05	5.24	0.58	1.00	7.4	22.6	0.75
180.00	3.14	2.00	3.14	6.28	0.50	0.00	6.7	21.0	NP

Example 10.2: A 200-ft concrete culvert is placed on a grade of 1.5 percent to carry a design flow of 200 cfs. Size a commercial pipe for this case.

Solution:

$$d = \left(\frac{NQ}{K\sqrt{S_0}}\right)^{\frac{3}{8}} = \left(\frac{0.014 \times 200}{0.462\sqrt{0.015}}\right)^{3/8} = 4.32\ ft = 51.8\ inches \quad \text{Use d=54 inches.}$$

Example 10.3: Recommend a commercial concrete pipe placed on a slope of 1.5% for a design flow of 40 cfs.

Solution: Apply Eq 10.13 to the five flow condition as:

$$d = \left(\frac{NQ}{K\sqrt{S_0}}\right)^{\frac{3}{8}} = \left(\frac{0014 \times 40}{0.462\sqrt{0.015}}\right)^{\frac{3}{8}} = 2.36\ ft \quad \text{Use d=2.5 ft = 30 inches}$$

By trial and error, the central angle for the design flow is found to be 115.23 degrees. Details are listed below:

A half Central Angle degree	Angle radian	Depth Y ft	Area A ft^2	Wprmtr P ft	H Radius R ft	Top Width T ft	Velocity V fps	Flow Q fps	Froude Fr	Remarks
115.23	2.01	1.78	3.74	5.03	0.74	2.26	10.7	40.0	1.46	Design Flow
180.00	3.14	2.50	4.91	7.85	0.62	0.00	9.5	46.6	NP	Full Flow

Example 10.4: Continue Example 10.3 to determine the critical flow condition.

Solution: By definition, the flow Froude Number in Equation 10.8 is set to be unity as:

$$F_r = \sqrt{\frac{Q^2T}{gA^3}} = 1.0$$

By trial and error, the central angle for the critical flow is found to be 143 degrees. Details are listed below:

Half Central Angle (0<theda-c<3.14)	theda-c=	2.35	radians or	134.72	degree
Critical flow area	Ac=	4.45	sq ft		
Critical top width	Tc=	1.78	ft		
Critical flow depth	Yc=	1.06	ft		
Critical flow velocity	Vc=	8.98	fps		
Froude Number	Fr=	1.00			

10.3. CULVERT HYDRAULICS

A culvert flow is controlled by either the orifice flow hydraulics at the entrance or the pipe flow through the culvert pipe, whichever requires a higher headwater depth for the given design flow. Therefore, the culvert flow must be analyzed for the two following conditions as:

(1) *inlet control culvert,* and
(2) *outlet control culvert.*

The headwater required by the orifice flow formula at the entrance is termed *inlet control headwater depth.* The headwater required by the energy balance between the entrance and the known tailwater depth is termed *outlet control headwater depth.* When the inlet control headwater depth is greater than the outlet control headwater depth, the culvert operates under inlet control; otherwise, it is under outlet control. A culvert may behave like inlet control to pass one discharge and then switches to outlet control for passing another discharge. Therefore, for a given discharge, the culvert must be examined by both the orifice formula and the energy grade line in order to know which one dictates the operation.

10.3.1. Culvert Hydraulics under Inlet Control

When a culvert is laid on a steep slope, the culvert capacity is controlled by the headwater depth at the entrance and the critical flow at the entrance. Such a culvert is termed inlet control culvert. As illustrated in Figure 10.5, the balance of flow energy between the approaching flow and the culvert entrance is written as:

$$H_1 = H_c + \frac{U_c^2}{2g} \quad \text{(if elevations are used in computation.)} \tag{10.14}$$

$$Y_1 = Y_c + \frac{U_c^2}{2g} + LS_o \quad \text{(if water depths are used.)} \tag{10.15}$$

$$U_c = \sqrt{2g(H_1 - H_c)} = C_o\sqrt{2g(Y_1 - LS_0 - \frac{d}{2})} \tag{10.16}$$

$$Q = C_o A\sqrt{2g(Y_1 - LS_o - \frac{d}{2})} = C_o A\sqrt{2g(H_w - \frac{d}{2})} \tag{10.17}$$

In which H= water surface elevation in [L], Y= water depth in [L] relative to the culvert exit invert, U= flow velocity in [L/T], L= length of culvert in [L], S_o= slope of culver pipe in [L/L], C_o= discharge coefficient 0.55 to 0.65, A= cross sectional area of culvert pipe in [L^2]. The subscript,1, means the variables associated with Section 1. Similarly, the subscript, c, represents the variables associated with the critical flow at the culvert entrance.

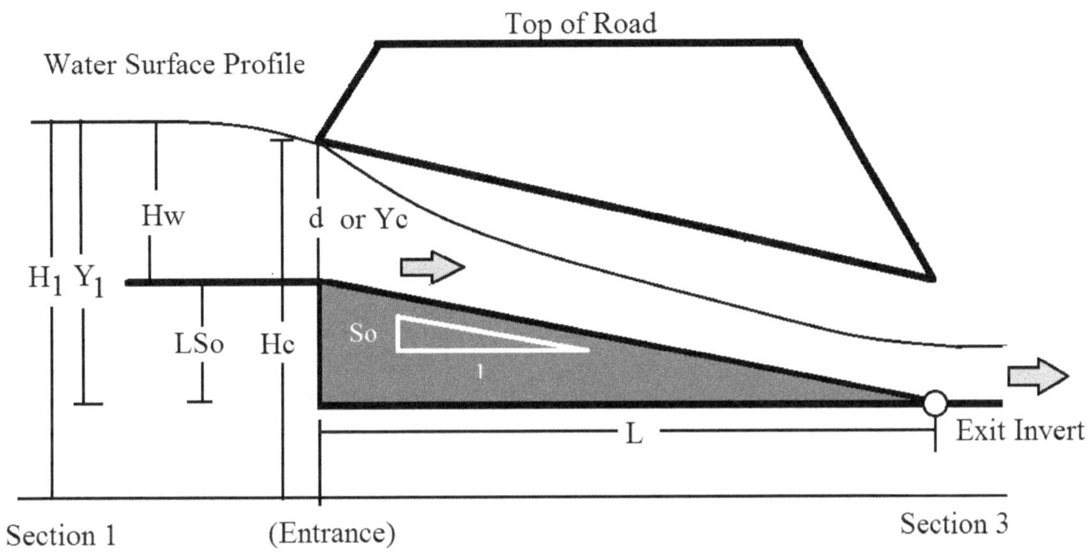

Figure 10.5. Culvert Hydraulics under Inlet Control.

For a given design flow, Q, Eq 10.17 defines the required inlet-control headwater depth, $H_{w\text{-}IN}$.

$$H_{W-IN} = H_w \quad \text{defined by Eq 10.17.} \tag{10.18}$$

As manifested in Eq 10.17, the capacity through an inlet-control culvert is independent of the culvert length, barrel slope and roughness.

10.3.2. Culvert Hydraulics under Outlet Control

For a flat culvert, the culvert hydraulics is dominated by the energy balance between the entrance, Section 1 in Figure 10.6, and the exit section. Obviously, the friction loss in this case depends on tailwater depth and culvert pipe's roughness, flow line slope, and culvert length.

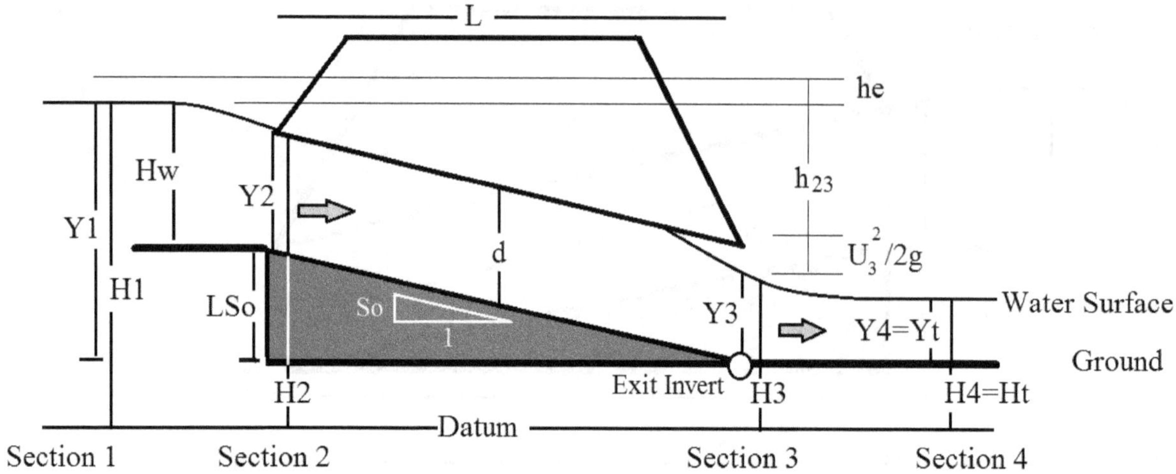

Figure 10.6. Culvert with Submerged Inlet and Outlet.

Applying the energy principle between Sections 1 and 3 in Figure 10.4 yields:

$$Y_1 = H_w + LS_o = Y_3 + \frac{U_3^{\ 2}}{2g} + h_e + h_{23} \tag{10.19}$$

Under the full flow condition $U_2 = U_3 = U_f$

$$h_e = K_e \frac{U_2^{\ 2}}{2g} \approx K_e \frac{U_f^2}{2g} \tag{10.20}$$

$$h_{23} = S_f L = \left[\frac{N^2 U_f^{\ 2}}{k^2 R^{\frac{4}{3}}} \right] L = \left[\frac{2g}{k^2} \frac{N^2}{R^{\frac{4}{3}}} L \right] \frac{U_f^{\ 2}}{2g} = K_N \frac{U_f^2}{2g} \tag{10.21}$$

$$Y_3 = \max[Y_t, \frac{Y_c + d}{2}] \tag{10.22}$$

$$Q = A \left[\frac{2g(Y_1 - Y_3)}{1 + K_e + K_N} \right]^{0.5} = A \sqrt{\frac{1}{1 + K_e + K_N}} \sqrt{2g(H_w + LS_o - Y_3)} \tag{10.23}$$

in which h_{23}= friction loss in [L] between Sections 2 and 3, U_f = full flow velocity in [L/T], K_e= entrance loss coefficient, Y_t= tail water depth in [L], Y_c= critical flow depth in [L], K_N= roughness coefficient, S_f= friction slope in [L/L], and k= 1.486 for feet-second units or 1.0 for meter-second units. The subscript, 1, represents the variables associated with Section 1. For a given design flow, Q, Eq 10.23 defines the required outlet-control headwater depth, $H_{w\text{-}OUT}$.

$$H_{w-OUT} = H_w \quad \text{defined by Eq 10.17.} \tag{10.24}$$

10.3.3. Determination of Culvert Capacity

As discussed before, for a given design flow, we shall analyze the culver hydraulics for both inlet and outlet control conditions. The design headwater depth is determined by:

$$H_w = \max(H_{w-IN}, H_{w-OUT}) \tag{10.25}$$

On the contrary, for a given headwater depth at the entrance, the culvert capacity is determined by the smaller one between Eq's 10.17 and 10.23.

Example 10.5: As shown in Figure 10.7, a discharge of 40 cfs is carried by a 30-inch circular culvert laid on a slope of 1.5%. The roughness coefficient of the sewer is 0.014. The water depth at the exit of the culvert is 2.0 feet. Determine the headwater depth for the design flow.

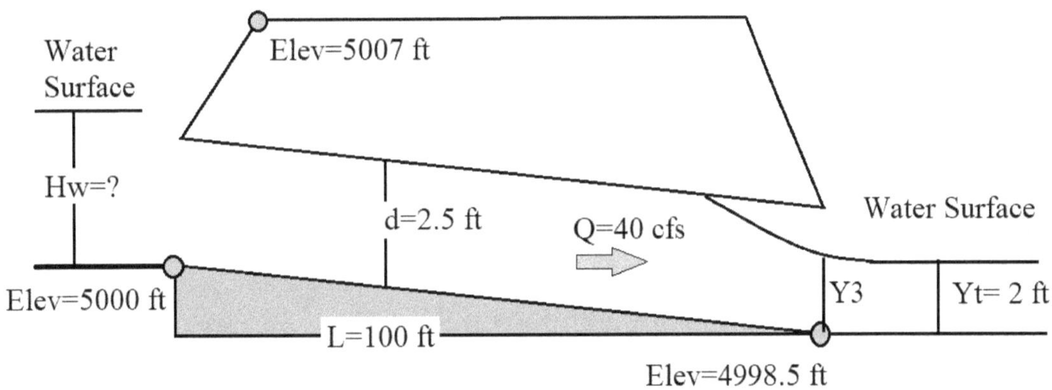

Figure 10.7. Design of Culvert.

Solution:

Design Information (input)			
Pipe Diameter in feet	d=	2.5	ft
Invert Elevation at Entrance	E-in=	5000.0	ft
Invert Elevation at Exit	E-out=	4998.5	ft
Pipe Length	L =	100.0	ft
Pipe Slope	S_o =	0.0150	ft/ft
Manning's Roughness	N =	0.0140	
Entrance Loss Coeff	K_e =	0.50	
Exit Loss Coeff	K_x =	0.00	
Bend Loss Coeff	K_b =	0.00	
Total Design Discharge	Qt=	40.00	cfs

Headwater Depth by Inlet Control			
Orifice Coefficient	Co =	0.52	
Headwater Depth by Inlet Control	Hw-IN	**5.06**	ft
Pipe Cross Sectional Area	Ao=	4.91	sq ft
Pipe Flow Velocity	Vo=	8.15	fps

Headwater Depth by Outlet Control			
If known, enter the Tailwater Depth	Yt =	2.00	ft
Y3=Max(Y-tail, 0,5(d+Yc))	Y3=	2.31	ft
Friction Loss Coeff	K_f =	1.06	
Sum of All Loss Coeff's	$K_{'s}$ =	1.56	
Headwater Depth by Outlet Control	Hw-OUT	**3.46**	ft

Design Headwater Depth Ratio			
Headwater for Design	HW=	**5.06**	ft
HW/d ratio =	HW/d=	**2.03**	

As recommended in Table 10.1, the acceptable ratio of Hw/d<1.5. Therefore, a pipe of 3.0 ft shall be used for this case. Repeat the above process, the solution for this case is: $Y_c = 2.06\ ft$, $H_{w\text{-}IN} = 3.34\ ft$, $Y_3 = 2.53\ ft$, $H_{w\text{-}OUT} = 2.19\ ft$, $H_w/d = 1.11$.

10.4. CLOSING

Culverts are often used as a drainage facility in urban areas. The major concerns associated with culvert design include, but not limited to, hydraulic efficiency and public safety. On top of the hydraulic analysis, it is recommended that a trash rack, as shown in Figure 10.8a, be installed at the entrance. The function of such a trash rack is not only for urban debris control, but also a life saver in case of any accident involving human or/and animal.

(a) Trash Rack at Entrance | (b) Fence for Safety

Figure 10.8. Safety around Culvert.

Fencing and railing are important elements in construction of large culvert structures. Figure 10.8b presents a case that needs more fences upstream of the culvert bridge. It is a good practice to fence around a culvert structure with sufficient flash flood warning signs.

10.5. HOMEWORK

Q10.1: A circular culvert of 36-inch in diameter is laid on a slope of 1.5% with Manning's N=0.025. Determine the flow capacity when H_w/d=0.75, 1.0, and 2.0.

Q10.2: A box culvert of 3-ft by 4-ft (height by width) is laid on a slope of 1.5% with Manning's N=0.025. Determine the flow capacity when H_w/d=0.75, 1.0, and 2.0 in which d=4 ft or the height of the box culvert.

Q10.3: A crossing culvert under a highway is designed to pass the design flow of 45 cfs through a length of 300 feet. A single concrete barrel of 3-ft in diameter is laid on a slope of 2.0%. The Manning's n for the barrel is 0.020. The entrance and exit loss coefficients are K_e=0.5 and K_X=1.0. The tailwater depth for the design condition is one foot above the crown at the exit.

(1) Determine the design headwater depth at the entrance of the culvert.

(2) The available water depth at the entrance is 8 feet. Can you use the head water of 8 feet to reduce the culvert size? If so, what is the diameter of the culvert? If not, explain.

Q10.4: As shown in Figure HW10.1, a culvert system is designed to pass the design flow of 260 cfs underneath the highway. Verify that two concrete pipes of 60 inches in diameter are acceptable for this case.

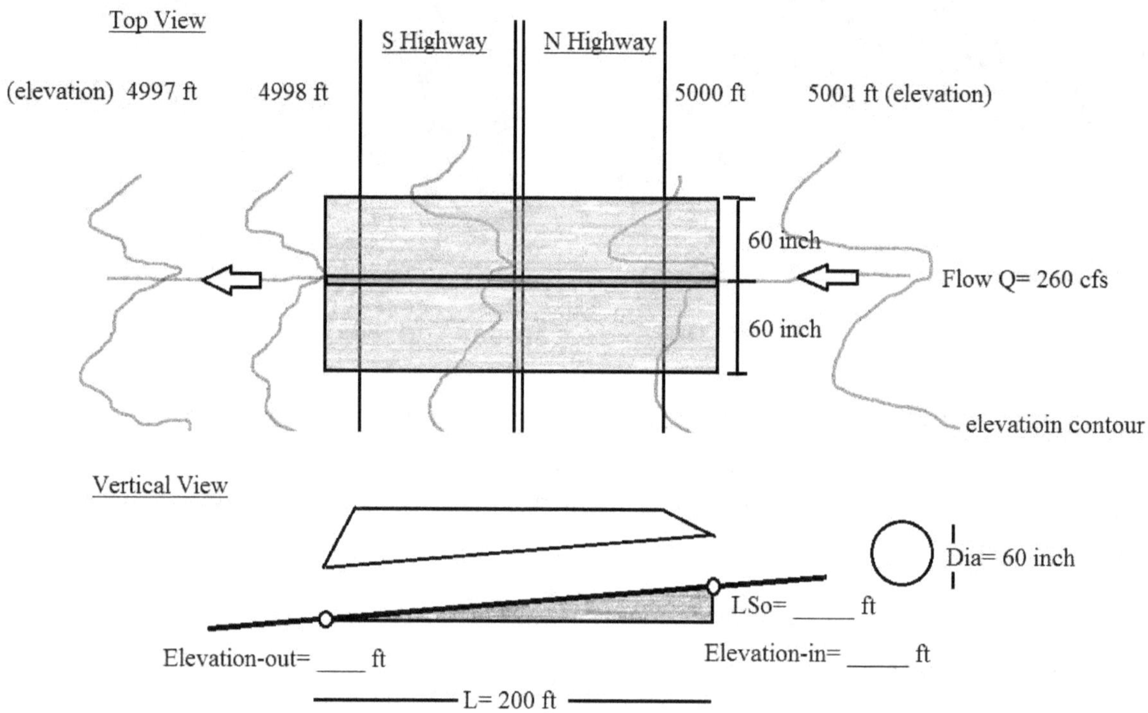

Figure HW10.4. Design of Culvert Bridge.

REFERENCES

Amein, M. and Fang, C.S., "*Implicit Flood Routing in Natural Channels*," ASCE J. of Hydraulic Engineering Division, Vol 96, Hy12, 1970, pp 24181-2500.

Arizona Department of Water Resources, *"Design Guideline for Diversion Channels"*, Flood Warning and Dam Safety Section, Arizona.

Bedient, P.B. and Huber, W.C., "Hydrology and Floodplain Analysis", 2nd Edition, Addison Wesley Publishing Company, New York, 1992.

Beuter, E.L., Gaebe, R.R. and Horton, R.E., "*Sprinkled-Plat Runoff and Infiltration Experiments on Arizona Desert Soils*", Trans, Amer. Geophys. Union, pp 550-558, 1940.

Brock, R.R., "Development of Roll Wave Trains in Open Channels," Proceedings of the ASCE J of the Hydraulics Division, Vol 95, No HY 4, July, 1969.

Chow, V. T., "*Open Channel Hydraulics*", McGraw-Hill Civil Engineering Series, New York, New York, 1959.

Department of the Army, "*Hydraulic Design of Flood Control Channels*", , Corp of Engineers, Washington DC, 20314-1000, EM 1110-2-1601.

Escoffier, F. F., and Boyd, M. B., "Stability Aspects of Flow in Open Channels", J. of the Hydraulics Division, ASCE, Vol 88, No HY 6, November, 1962.

Fread, D.L. "*National Weather Service Operation Dynamic Wave Model,*" Verification Math. and Physical Models in Hydraulic Engr., Proc. 26th Annual Hydr., Div., Special Conference, ASCE, College Park, Maryland, 1978.

French, Richart H., "Open Channel Hydraulics", McGraw Hill, New York, New York, 1985.

Eagleson, P.S., "*Dynamic Hydrology*", McGraw Hill, New York, pp 344-346, 1970.

Ippen, Arthur T., "*Mechanics of supercritical flow*", Transactions, ASCE, VOL 116, pp 268-295, 1951.

Knapp, Robert T., "*Design of Channel Curves for Supercritical Flow*", ASCE Transaction, November 1949, pp 296-325.

Guo, James C.Y., and Hughes, William, "*Optimal Trapezoidal Cross Sections for Open Channel with Freeboard*", American Society of Civil Engineers' Journal of Irrigation and Drainage Engineering, Vol 110, No. 10, October 1994.

Guo, James C.Y. "*Overland Flow on a Pervious Surface*", IWRA J. of International Water, Vol 23, 1998.

Guo, James C.Y., "*Roll Waves in High Gradient Channels,*" International Water Resources Association's International Journal of Water , Vol 3, No.1, March 1999.

Guo, James C.Y., Section 700, *"Open Channels*", Hydrologic Criteria and Drainage Design Manual", published by Clark County Regional Flood Control District, Clark County, Las Vegas, NV, 1999.

Guo, James C.Y, *"Urban Flood Control and Stormwater Management*." CSC press, New York, 2017.

Guo, James C.Y. "*Critical Flow Section in a Collector Channel,*" ASCE J. of Hydraulic Engineering, Vol 125, No. 4, April, 1999.

Guo, James C.Y. "*Risk-cost Approach to Interim Drainage Structure Designs,*" ASCE J. of Water Resources Planning and Management, Vol 124, No 6, Nov/Dec, 1998.

Guo, James C.Y. "*Design of Drop Structure*", J. of Flood Hazards News, Urban Drainage and Flood Control District, Denver, Colorado, Dec. 2004.

Guo, James C.Y. "*Maximal Dissipation for Drop Structure Design*", reviewed and accepted for publication by ASCE J. of Hydraulic Engineering, October, 2004.

Henderson, F.M., and Wooding, R.A., "*Overland Flow and Groundwater from a Steady Rainfall of Finite Duration*," J. Geophys. Res., Vol 69, No 8., 1964, pp 39-67.

Holden, A.P. and Henson, D., "*Finite Difference Formulations of Kinematic Equations*", J. of Hydraulic Engineering, ASCE, pp 423-426, May, 1995.

Holden, A.P. and Henson, D., "*Improved four-point Solution of the Kinematic Equations.*", J. of Hydro. Res, 26(4), 1988, pp 413-423.

Horton, R.E., "*The Interpretation and Application of Runoff Plot Experiments with Reference to Soil Erosion Problems*," Proceeding, Coil Science Society of America, Vol 3, 1938, pp 340-349.

Hydrologic Engineering Center, "*Introduction and Application of Kinematic Wave Routing Techniques Using HEC-1*," Training Document 10, Unite State Army Corps of Engineers, Davis, California.

Izzard, C.F. "*The Surface Profile of Overland Flow*," Transactions, American Geophysical Union, Vol 25, No. 6, 1944, pp 959-968.

Izzard, C.F. "*Hydraulics of Runoff from Developed Surfaces*," Proceedings, Highway Research Board, Washington, D.C., Vol 26, 1946, pp 129-146.

Katz, D. M., Watts, F. J., and Burroughs E.D., "*Effects of Surface Roughness and Rainfall Impact on Overland Flow*", J. of Hydraulic Engineering, ASCE, Vol 121, No 7, July 1995.

Liggett, James A. "Stability", Chapter 6, Unsteady Flow in Open Channels, Edited by Mahmood, K., and Yevjevich, V., Vol 1, Water Resources Publications, Littleton, Colorado, 1975.

Lighthill, M.H. and Whitham, G.B., "*On Kinematic Waves, I, Flood Movement in Long Rivers*", Proc., Royal Soc., of London, Ser. A., Vol 229, pp 281-316, 1955

Mayer, P.G.W. "A study of Roll Waves and Slug Flows in Inclined Open Channels", PhD Dissertation, Cornell University, 1957

McCuen, R. H., Wong, S. L., and Rawls W. J., "*Estimating Urban Time of Concentration*", J. of Hydraulic Engineering, ASCE, Vol 110, No. 7, July 1984.

Morgali, J.R., "*Laminar and Turbulent Overland Flow Hydrographs*", J. of Hydraulic Engineering, ASCE, HY 2, pp 441-360, 1970

Morgali, J.R., and Linseley, R.K., "*Computer Analysis of Overland Flow*", J of Hydraulic Engineering, ASCE, HY 3, pp 81-100, 1965.

Overton, D.E. and Meadows, M.E. "*Stormwater Modeling,*" Academic Press, New York. 1976

Prima County, "*Drainage and Channel Design Standards for Local Drainage*", Department of Transportation and Flood Control District, 1984

Preissmann, A. "*Propogation des intumescences dans les canaux et les riveres*." I'e Congres de I'association Francaise de Culcule, Grenoble, France, 1961, pp433-442.

Ponce, V.M., and Yevjevich, V. "*Muskingum-Cunge Method with Variable Parameters.*" ASCE J. of Hydr. Divi., 104(2), 1978, pp1663-1667.

Robert Manning, "*On the flow of water in open channel and pipes*", Transactions, Institution of Civil Engineers of Ireland, Vol 20, pp161-207, Dublin, 1891

Rouse, Hunter Rouse, B. Hoota, B.V., and Hsu, Eq 9En-Yun, "*Design of Channel Expansions*", ASCE Transactions, November 1949, pp 347-400

Shunkry, Ahmed, "*Flow Around Bends in an Open Flume*", ASCE Transactions, Paper No. 2411, Vol 115, pp 751-779, 1950

Schaake, Jr., Geyer, J.C., and Knapp, J.W., "*Experimental Examination of the Rational Method*: ASCE J. of Hydraulic Engineering, Vol 93, HY6, 1967, pp 353-370.

Singh, V.P. and Cruise, J.F., "*Analysis of the Rational Formula Using a System Approach*", Catchment Runoff and Rational Formula, edited by B.C. Yen, pp 39-51, Water Resources Publication, Colorado, 1992

Stonstreet, Scott E., "Stochastic Determination of Wave Heights for Flood Control Channels", ASCE Conference Proceedings of North American Water and Environment Congress, 1996

Thorsky, G. N., and Haggman, D.C., Discussion of "Development of Roll Wave Trains in Open Channels," ASCE J of Hydraulic Division, Vol 96, No HY 4, April, 1970.

Urban Highway Storm Drainage Model, "*Inlet Design Program*", Vol 3, Federal Highway Administration, Report No. FHWA/RD-83/043, December 1983.

US Army Corps of Engineers, "Hydraulic Design of Flood Control Channels," Engineer Manual No. 1110-2-1601, Department of the Army, Washington D.C., 20314-1000, 1991

USGS Open File Report 82-873, "*Rainfall-runoff Data from Small Watersheds in Colorado, October 1977 through September 1980*", USGS, Lakewood, Colorado, 1983

Urban Drainage and Flood Control District (2010) "*Urban Drainage and Stormwater Design Criteria*", Vol 1 and 2, Denver, Colorado.

Vedernikov, V. V., "Conditions at the Front of a Translatory Wave Distributing a Steady Motion of a Real Fluid", C.R. USSR Acad. Sci., Vol 48, No 4, 1945.

Vedernikov, V.V., "Characteristic Features of a Liquid Flow in an Open Channel", C. R. USSR Acad. Sci., Vol 52, 1946.

Woolhiser, D.A., and Liggett, J.A., "*Unsteady One-dimensional Flow over a Plane - the Rising Hydrograph*", Water Resources, Vol 3, No 3, pp 753-771, 1967

Wooding, R.A, "*A Hydraulic Model for a Catchment-Stream Problem*", J. of Hydrology, Vol 3, pp 254-267, 1965

Yen, B.C. and Chow, V.T., "*Experimental Investigation of Watershed Surface Runoff*", Hydraulic Engineering Series, No 29, Dept of Civil Engineering, U of Illinois at Urbana- Champaign, September 1974.

www.ingramcontent.com/pod-product-compliance
Lightning Source LLC
LaVergne TN
LVHW082341200926
848481LV00012B/3415